Transonic Compressor Stages with Non-Axisymmetric End Walls

Dr. Steffen Reising

Transonic Compressor Stages with Non-Axisymmetric End Walls

Unsteady and Steady Performance

Südwestdeutscher Verlag für Hochschulschriften

Impressum/Imprint (nur für Deutschland/only for Germany)
Bibliografische Information der Deutschen Nationalbibliothek: Die Deutsche Nationalbibliothek verzeichnet diese Publikation in der Deutschen Nationalbibliografie; detaillierte bibliografische Daten sind im Internet über http://dnb.d-nb.de abrufbar.
Alle in diesem Buch genannten Marken und Produktnamen unterliegen warenzeichen-, marken- oder patentrechtlichem Schutz bzw. sind Warenzeichen oder eingetragene Warenzeichen der jeweiligen Inhaber. Die Wiedergabe von Marken, Produktnamen, Gebrauchsnamen, Handelsnamen, Warenbezeichnungen u.s.w. in diesem Werk berechtigt auch ohne besondere Kennzeichnung nicht zu der Annahme, dass solche Namen im Sinne der Warenzeichen- und Markenschutzgesetzgebung als frei zu betrachten wären und daher von jedermann benutzt werden dürften.

Verlag: Südwestdeutscher Verlag für Hochschulschriften GmbH & Co. KG
Dudweiler Landstr. 99, 66123 Saarbrücken, Deutschland
Telefon +49 681 37 20 271-1, Telefax +49 681 37 20 271-0
Email: info@svh-verlag.de

Approved by: Darmstadt, TU, Diss., 2011

Herstellung in Deutschland:
Schaltungsdienst Lange o.H.G., Berlin
Books on Demand GmbH, Norderstedt
Reha GmbH, Saarbrücken
Amazon Distribution GmbH, Leipzig
ISBN: 978-3-8381-2737-8

Imprint (only for USA, GB)
Bibliographic information published by the Deutsche Nationalbibliothek: The Deutsche Nationalbibliothek lists this publication in the Deutsche Nationalbibliografie; detailed bibliographic data are available in the Internet at http://dnb.d-nb.de.
Any brand names and product names mentioned in this book are subject to trademark, brand or patent protection and are trademarks or registered trademarks of their respective holders. The use of brand names, product names, common names, trade names, product descriptions etc. even without a particular marking in this works is in no way to be construed to mean that such names may be regarded as unrestricted in respect of trademark and brand protection legislation and could thus be used by anyone.

Publisher: Südwestdeutscher Verlag für Hochschulschriften GmbH & Co. KG
Dudweiler Landstr. 99, 66123 Saarbrücken, Germany
Phone +49 681 37 20 271-1, Fax +49 681 37 20 271-0
Email: info@svh-verlag.de

Printed in the U.S.A.
Printed in the U.K. by (see last page)
ISBN: 978-3-8381-2737-8

Contents

List of Figures

List of Tables

Nomenclature

Roman Symbols

A	$[m^2]$	area
a	$[ms^{-1}]$	speed of sound
c	$[m]$	chord
c_p	$[kJkg^{-1}K^{-1}]$	specific heat capacity at constant pressure
CP	$[-]$	pressure loss coefficient
c	$[ms^{-1}]$	absolute velocity
f	$[Hz]$	frequency
D		diffusion factor
f,k	$[ms^{-2}]$	specific volume force
h	$[kJkg^{-1}K^{-1}]$	enthalpy
h	$[m]$	blade height
l	$[m]$	length
M	$[-]$	Mach number
m	$[kgs^{-1}]$	mass flow rate
P		Penalty
p	$[Pa]$	pressure
R	$[kJkg^{-1}K^{-1}]$	gas constant
r	$[m]$	radius
$\dot{S}$	WK^{-1}	entropy flux
S	$[m^2]$	Surface
s	$[m]$	blade pitch
s	$[kj/kg^{-1}K^{-1}]$	specific entropy
t	$[s]$	time
T	$[K]$	temperature
u	$[ms^{-1}]$	blade rotational speed
V		Value
V	$[m^3]$	Volume
v	$[ms^{-1}]$	absolute velocitdy
w	$[ms^{-1}]$	relative velocity
w	$[-]$	weighting factor
x,φ,r	$[m,°,m]$	polar coordinates
x,y,z	$[m,m,m]$	cartesian coordinates

Greek Symbols

α	$[°]$	flow angle from the axial direction

η	[-]	efficiency
θ	[m]	momentum thickness of wake or boundary layer
κ	[-]	isentropic coefficient
λ	[m]	wavelength
μ	$[kgm^{-1}s^{-1}]$	dynamic viscosity
μ_t	$[kgm^{-1}s^{-1}]$	turbulent viscosity
ν	$[m^2s^{-1}]$	kinematic viscosity
ρ	$[kgm^{-3}]$	fluid density
σ	[]	solidity
τ	$[Nm^{-2}]$	shear stress
Ω	$[s^{-1}]$	angular velocity
ω	$[s^{-1}]$	angular velocity
ω	[-]	pressure loss coefficient
ξ	[-]	pressure loss coefficient
Π	[-]	total pressure ratio
Φ		scalar parameter

Subscripts

abs	absolute
ax	axial
dyn	dynamic
DP	design point
EW	Euler walls
in	inlet
NS	Navier Stokes
norm	normalized
out	outlet
pol	polar coordinates
prim	primary
red	reduced
ref	reference
rel	relaive
req	required
sec	secondary
sta	static
tot	total
1	inlet
2	outlet

Acronyms

| ANN | Artificial Neural Network |

BPF	Blade Passing Frequency
BLISK	Bladed Disk
CAD	Computer Aided Design
CFD	Computational Fluid Dynamics
CFL	Courant Friedrichs Lewy
GA	Genetic Algorithm
HP	High Pressure
IGV	Inlet Guide Vane
IP	Intermediate Pressure
LP	Low Pressure
LE	Leading Edge
LER	Leading Edge Recambering
NGV	Nozzle Guide Vane
OF	Objective Function
OGV	Outlet Guide Vane
PEW	Profiled End Walls
PS	Pressure Side
SS	Suction Side
RANS	Reynolds Averaged Navier Stokes
rpm	revolutions per minute
RSI	Rotor Stator Interface
SKE	Secondary Kinetic Energy
SKEH	Secondary Kinetic Energy weighted with Helicity
TAVG	Time-Averaged
TE	Trailing Edge
TPR	Total Pressure Ratio

1 Introduction

The progressive globalization and the growing world population have led to an increased demand for fossil fuels. The worldwide resources of these fuels therefore become more and more expensive and precious. Until the year 2030, a rise in energy consumption by 50% is forecasted. In the year 2060, this additional need is even assumed to have grown up to 200%. The main reasons for this development lie in the growth of the industrial sector and in the fact that the economic status of the population in the emerging countries, in particular the BRIC-countries (Brazil, Russia, India and China), will probably have adapted to the standard of the western industrial nations [55]. This drives the air traffic to gain a higher importance in the context of which an increase of 5% per year is expected. Since this demand is in conflict to the global concern for resource preservation and reduction of energy consumption, jet engine design technology needs to be improved in terms of higher efficiency, less emissions, less weight and also considering alternative engine concepts. However, it has become quite difficult to further improve theses targets as the design of modern jet engines is already very sophisticated. In 2001, the report 'European Aeronautics - a vision for 2020' was published. This report had been composed by representatives of the important European aviation companies, amongst others, defining three major goals for the year 2020:

- Reduction of the fuel consumption and CO_2-emissions by 50%

- A 50% reduction of the noticeable noise

- Reduction of the NO_x-emissions by 80%

The design process of an aero engine is a very complex and time-consuming task which is driven by many different objectives and constraints. Until the 80s, this was especially challenging as almost the entire design had to be done analytically by the engineers. Today, the overall design process is arranged in component-based subtasks where different design tools and disciplines are involved to fulfill the design requirements [56]. Hence, the design quality of every single engine component (i.e. compressor, combustion chamber, turbine (Figure 1.1)) impacts the overall engine performance in either positive or negative ways.

The efficiency of turbomachinary components in today's jet engines and also stationary heavy-duty gas turbines is already so high that the objective is often to keep the level of efficiency (or to improve it only slightly) while increasing the aerodynamic load at the same time. This leads to a reduced number of compressor/turbine stages and as a consequence to smaller and lighter engines including an economy in life-cycle costs.

1.1 Compressor Design

Compressor design is a very challenging, highly complex and time consuming task involving many different constraints and requirements. Because compressor blade rows feature diffusing flow, everything is connected to everything else. Blockage on the end wall, such as separation,

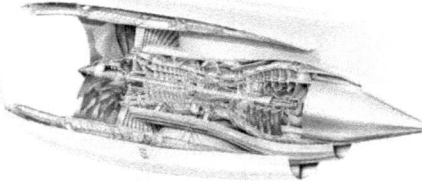

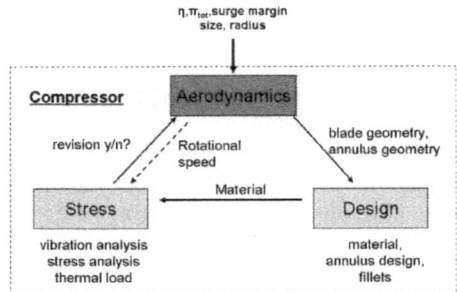

Figure 1.1: Major components of a jet engine from [78] [Printed by courtesy of Rolls-Royce Deutschland]

Figure 1.2: Iterative multi-disciplinary compressor design process [15]

e.g. affects the whole flow field and also quite perversely, even though in itself it is undesirable, it can still have beneficial effects in terms of stability. In recent decades, the development of axial aircraft compressors has led to extremely high stage loadings and therewith reductions in entire engine size and weight. This trend has several disadvantages, such as the risks of flow separation (and thus of reduced surge margin) and higher secondary flows (and thus reduced efficiency) that are associated with increased stage loading. High-lift airfoils may therefore generate secondary losses that negate the weight and cost benefits. This impact was e.g. shown by the work of Haselbach et al. [41] who applied high-lift airfoils to the Rolls Royce BR715 LP Turbine. Rig testing resulted in a 0.5% decrease in stage efficiency which the authors ascribed to the higher end wall losses and pressure side separation bubbles that arose from their high-lift blade design. In this regard, compressor design can be frustrating because simply making one feature better may be outweighted by something else getting worse as a result.

This is of high criticality due to the compressor's importance within an jet engine. The compressor is responsible for approximately 50% of the entire engine length and weight as well as for 40% of the manufacturing costs [90]. The multi-disciplinary compressor design process is preceded by the preliminary design of the whole jet engine. A conceptional study according to the market requirements is carried out. This study must consider the exclusion criteria determined by the airframer such as thrust, weight, costs and specific fuel consumption. Often, the flight cycle as a secondary criterion is also considered. For this purpose, correlations based on already existing engines are used. In this design phase, the responsible department has to communicate with the compressor group concerning the availability of the range of products. Once, the new engine is contracted, the thermodynamic cycle is composed where not only the main operating point of the flight cycle (cruise) but also part load (idle, approach) and over load (take-off, climb) behavior are taken into account. The reference values for all components will be defined through this analysis.

The following compressor design process itself consists of several separate design disciplines, see Figure 1.2. The iterative, sequential design procedure starts with the aerodynamic layout. In this initial step, the velocity triangles on the compressor mid-height line and the annulus geometry are defined by means of simple one-dimensional calculations. Many other important flow parameters, such as flow angles and velocities but also temperature and pressure values, can be derived from theses velocity triangles. Since this kind of calculation is of low time effort, it is efficient also to predict off-design characteristics of the compressor at this early stage. This meanline prediction is followed by a throughflow calculation. Here, the initial annulus geometry is smoothed and secondary features, such as cavities at the hub or casing, are eventually added. Moreover, the velocity triangles across the entire channel height and therewidth the corresponding flow angles can be obtained. With their help, the aerodynamic engineer can perform the 2D-blading process and stack the resulting blade sections radially afterwards. Concerning state-of-the-art compressor design, 3D-blading is then performed by introducing design principles as sweep and dihedral in order to minimize secondary losses [30]. In the subsequent design step, the material is defined and the blades are aligned in the annulus. Furthermore, the fillet radii between the blade and end wall surfaces are added. A stress analysis has then to proof whether the blade geometry with the used material can withstand the assumed loads or not. In this context, conflicts may appear between aerodynamics and stress. For the aerodynamic layout the blade geometry is optimized regarding 'cruise' (i.e. operating point in which jet engine remains most of the time during a flight cycle). However, the loss in efficiency should not be unacceptably high at off-design conditions and a sufficient surge margin must be verified in the compressor map. Yet, when it comes to strength, the most critical operating point is take-off due to the highest rotational speeds resulting in the highest mechanical loads. Here, a rather big cross section of the blade would be of advantage which, in turn, would have a negative impact on the aerodynamic efficiency. The aerodynamic compressor design process is therefore not straight forward but requires many loops. If the design targets can not be achieved in one of the design disciplines modifications on design parameters and assumptions have to be made. This can also mean that the designers have to step back to one of the previous design loops. In order to save time, as many processes as possible are intended to be set up simultaneously. Whilst the aerodynamic design is not finished yet, a dummy stress model can already be established based on a similar blade that is part of the product database. The new blade, once the aerodynamic design is done, can then quickly be integrated in the stress model. The stress group can also modify the blade design within small tolerances and check their measures with a simple CFD-tool. On the other hand, several stress criteria are already considered during the aerodynamics in order to reduce the inner design loops.

After many time-consuming inner iterations among these three different disciplines, the best compressor design, being always a compromise, is hopefully found which satisfies all objectives and requirements of the design task. The compressor is introduced to the manufacturing process and finally integrated with all other components into the jet engine.

1.2 Modern Design Tools

Nowadays, due to the increased level of competition imposed by the market, the turbomachinery manufactures are forced to rapidly adjust their machines to the continuously changing market requirements. These demanding performance goals are carried out by increasing the efficiency, the work and pressure ratio per stage and the operating temperature while reducing the number of blades and stages and ensuring the mechanical integrity of the engine. Regarding the described time-consuming design process, the goal must be to reduce the number of design loops and shortening the entire design cycle by preventing human intervention from becoming a bottleneck, preserving the huge human experience gained during several decades. To achieve this goal, a more detailed understanding of the highly complex flow processes in the blade rows of turbomachinery and a consequent consideration of these findings within the applied design techniques are required. Basically, there are three different branches of fluid dynamics available for the investigation of fluid mechanical problems. In addition to the classical approaches as the analytical and experimental fluid dynamics, the computational fluid dynamics (CFD) has become firmly established in the last decades.

The analytical description of complex flows, as they appear in jet engine components, is only possible in a very limited field and only under simplifications concerning the physical reality. In order to gain a deeper understanding, just the experimental and numerical investigations of flow structures provide sufficient possibilities. However, experimental analysis on real or modeled designs are generally very expensive and laborious. It is quite challenging to guarantee the geometric, Mach and Reynolds analogies due to the thermodynamic boundary conditions which are difficult to maintain. An additional problem is the resolution of the used probes. Flow phenomena which lie under the resolution will be influenced by the probe or not even captured by the experiment.

By contrast, numerical investigations allow parameter variations with comparatively little additional effort. Characteristic flow parameters and analogies can be simulated in various combinations and also the thermodynamic boundary conditions are very flexible. In comparison to the experiments, the results of CFD simulations can be obtained in a shorter time and with less financial effort. The cost of a CFD calculation corresponded to 20% of a comparable experiment in the mid-90s [8]. On the continuous progress towards design methods satisfying the mentioned performance goals, one of the most important step was done at the end of the 80s by employing CFD and structural analysis codes in industry. Those codes have first been used in a trial and error procedure driven by an experienced designer [72]. Computational power has increased enormously over the last decades helping the designer to define advanced blade geometries, compute the flow field inside the blade channels and the mechanical stresses in the solid parts of the machine. Hence, this proportion has shifted even more in favor of the numerics as a further increasing personnel and instrumental expenditure faces decreasing numerical costs and more efficient codes. Typically, computing power has increased by more than a factor of 10^6 while gains in algorithmic performance can be estimated as close to 10^3 over the last 30 years [48]. As a consequence, full 3D turbulent Navier-Stokes multistage computations can be performed on a daily base in industry today.

The higher resolution of the numerical simulations provides the engineer a deeper insight into flow phenomena which are below the probe resolution due to their small length scales. The designer is also allowed to evaluate flow parameters in sections where it is simply not possible to gain access in the experiment. These two aspects connected with the flexibility when changing the thermodynamic and geometric boundary conditions provide the possibility to carry out numerical analysis in a rather simply and cheap manner.

Even considering the still existing shortcomings of modeling certain physical coherences, such as turbulence, the CFD, nonetheless, represents an enormous gain in knowledge and understanding the occurring flow phenomena. However, one has to admit that in particular the weaknesses of the used turbulence models can result in significant errors of the simulations. For this reason, a reasonable prediction of losses and their distribution is still task that needs to be improved. This holds also true for the capture profile and boundary layers which are dominated by laminar-turbulent transition [45].

The validation of numerical simulations of turbomachinery components can only be checked in comparison with experimental data. In this context, pressure distributions along solid surfaces (which are frequently used as comparison criterion) do not stay abreast the large amount of information provided by CFD simulations. More meaningful comparisons also refer to averaged or local loss and flow angle distributions as well as the correct shock position and the expansion of the blade wake. Therefore, the experiment remains an irreplaceable tool for the flow analysis.

1.2.1 Optimization Tools

Although considerable gains in performance have been made by the described approach, a further improvement will most likely be more difficult to achieve due to the increasing complexity of the flow problems to be investigated. The objective of the advancement of numerical codes must be to decrease the committed error of arbitrarily complex configurations to a minimum. The costly experiments would then only be needed for validation purpose. Even though CFD software is getting more accurate, fast and user-friendly it does not provide algorithms able to automatically modify a new geometry and predict its performance. As a consequence, blade designers often start from an existing geometry and try to adapt and improve it based on trial and error procedures. However, the request for very short design time schedule, does not allow the designers to test many modifications and therefore one cannot take full advantage of the huge potential and amount of information provided by the CAD, CFD and structural codes. I.e. although this method has already decreased the design time and costs thanks to a fewer number of experimental testing, human intervention is still demanded to drive the design process and therefore does not entirely satisfy the wish to minimize the design time and costs. Critical issues for the future require the need to incorporate CFD simulations, up to the level of 'fast prototyping', into the blade design optimization systems which calls for reliable and fast solvers for a very short process turnaround time. The advanced use of virtual prototyping and in particular advanced CFD tools will be fundamental in order to gain a competitive edge due to the high efficiency level of the current aero engines.

In this regard, the aspect of aerodynamic optimization has entered the turbomachinery design process in the recent past. Today, numerous design techniques, such as gradient methods based on finite differences, genetic algorithms, simulated annealing, response surface, random walk and others are available. Basically, they all can be used for the development of aerodynamic optimization in turbomachinery in each of the different steps of the design process [56], whereas it is hard to state, if not impossible, the superiority of one method over the others for any kind of design problem. Most times, these design problems are of a multi-objective character with conflicting design goals. As a further step, these optimization techniques are more and more in the process of being directly connected to 3D-CFD solvers in order to set up fully automated 3D-process chains. This is necessary as the technology moves progressively towards the overall application of multi-disciplinary optimization (MDO) [48]. Unfortunately, some of the mentioned optimization techniques require the evaluation of hundreds or even thousands of modified designs in order to predict a new geometry that is expected to be the optimum in performance of the entire design space. An important step to handle this drawback was the introduction of artificial neural networks (ANN) combined with a genetic algorithm which was presented by Pierret et al. [72] among the firsts. The ANN represents an approximated mathematical model of the real flow problem. This enables the performance evaluation of a new sample without running the proper CFD simulation but by calling a simple mathematical function. By means of the ANN the genetic algorithm can be used as an optimization technique in a very efficient manner. All these developments have led to an enormous speed-up of the turbomachinery design process by reducing both the interventions of the engineers and the computational efforts. Thus, an increasing importance of that research field is expected for the nearer future.

1.3 Content, Structure and Goals of this Study

In order to satisfy the rising demands on efficiency of future aero engines, knowledge about unsteady effects and their influence on the engine performance is mandatory. This is of particular importance since these unsteady effects have a big impact on the interaction between different components. Examples for such interactions are the clocking effect between combustion chamber and HP turbine and the variation of the combustion swirl in sense of rotation and strength, both having a major influence on the required cooling design of the first turbine stages, in particular the inlet guide vane. But also within a single component, unsteady effects should not be neglected. In today's aero engines and heavy-duty gas turbines the front stages of the compressor are transonic, an aspect, which involves rather complex shock systems. Considering the second stage, these shocks travel upstream into the stator of the first stage where they are reflected at the blade [17] influencing the static pressure recovery. They might even reach the rotor of the front stage. Therefore, one has increasingly tried to consider and focus on these effects over the last years.

On account of that, the *Institute of Gas Turbines and Aerospace Propulsion, the Institute for Energy and Power Plant Technology, the Institute for Numerical Methods in Mechanical Engineering* and the *Institute for Mechatronic Systems in Mechanical Engineering* established the post graduate program *Unsteady System Modelling of Aircraft Engines* at the *Technische Universität Darmstadt*. This research training group has been organized in close collaboration of *Rolls-Royce Deutschland* to allow the industrial needs. The interdisciplinary scientific objective of the research projects is the development of advanced methods, models and technologies to describe aero engines as a complete system considering that it is of the utmost importance to deal with the time-dependent analysis of flow phenomena by experimental and numerical means. To achieve the set targets, it is appropriate to arrange the research projects into the groups *model development for unsteady problems, validation, numerical methods* and *application*. The topic of the present study is incorporated in the application related sub-project *Flow Inhomogeneities in Compressors*. These inhomogeneities typically include secondary flow, separation, wakes, shock systems and rotor-stator interaction. The control and understanding of such phenomena is still an open technological question and of high priority for the industry which also highlights the interface to the industrial collaboration.

The emphasis of this work is to analyze the steady and unsteady performance of a transonic compressor stage with non-axisymmetric end walls where the profiled end wall serves to control the addressed flow inhomogeneities. The axisymmetric layout of *Configuration I* of the *Darmstadt Transonic Compressor* serves as the datum design. As a tool to find the optimum non-axisymmetric end wall shape, a fully-automated multi-objective optimizer connected to a steady 3D-RANS flow solver is used. The goal is to analyze how effective such a design tool can work on such a challenging task and to derive first design rules and compare the differences and features in common to the experience made by turbine researchers. In this context, the stator and rotor rows of the stage are subsequently and individually optimized. As a further step, the obtained optimized geometries are investigated in unsteady mode. Amongst others, two important questions shall be answered:

- *Does the delta in efficiency gained through the steady optimization match the change in performance when looking at the datum and optimized designs in unsteady mode?*

- *Do unsteady phenomena which cannot be captured within the optimization have any negative influence on the flow characteristics of the optimized design, i.e. would the optimized design have to be different if unsteady influences were, however this could be done, considered during the optimization?*

If the latter should be vitally important, this of course would conflict with the need to achieve shorter design cycles in the industrial environment.

This thesis is subdivided into seven chapters. Following the introduction, Chapter 2 gives an overview of the theoretical background regarding the most important aspects of this work. It contains an introduction both to the field of end wall profiling and the use of sweep and dihedral for 3D-bladings since both features have already proofed to control end wall flow. In chapter 3 the aspects of CFD will be discussed including the governing equations and unsteady treatment of flows in turbomachinery. Furthermore, this chapter describes the applied numerical optimization process chain. The following two chapters represent the design exercises on which the process chain is applied, starting with the stator optimization in Chapter 4, followed by the rotor optimization in Chapter 5. Chapter 6 deals with the unsteady simulations of the datum and optimized stator designs. Finally, this thesis will be completed with Chapter 7 providing conclusions and an outlook for future works in the field of end wall profiling in transonic compressors for jet engines.

2 Theoretical Background

The design of modern efficient compressors requires a global understanding of the three-dimensional flow structures inside the blading and the annulus geometry. This chapter introduces an overview of the most important physical aspects of this work. It contains an introduction to the field of secondary flows in turbomachinery. To describe and quantify secondary flows, a definition of a primary flow is requested. Throughout this study, two different definitions have been used which will both be presented afterwards. Then, an overview of the application of non-axisymmetric end walls is given being one method to influence secondary flow. Often, sweep and dihedral for 3D-bladings are used to control end wall flow. Therefore, it is worthwhile to give a short review of state-of-the-art 3D compressor airfoil design identifying the key effects and their benefits. An additional subchapter is dedicated to leading-edge modification and end wall fences which can also be interpreted as a sort of non-axisymmetric contouring although they are restricted to only a limited part of the passage. All the above mentioned features belong to the passive methods for controlling boundary layer and secondary flow. For the sake of brevity, active methods for flow control such as end wall boundary layer removal or flow injection will not be discussed. For the interested reader, the work of Gümmer et al. [31] of numerical investigations concerning boundary layer removal in highly-loaded axial compressor blade row is suggested as a source to this topic which also provides further readings and references.

2.1 Secondary Flow in Turbomachines

In recent decades the development of axial aircraft compressors has led to extremely high stage loadings and therewith reductions in entire engine size and weight. Secondary flows involving cross flow and three-dimensional separation phenomena in modern axial compressors at high stage loading contribute significantly to a reduction in overall efficiency and surge margin. The basic principles of the secondary flow mechanisms shall be outlined using a turbine as an example as the basic features of compressor secondary flows are the same. A schematic is illustrated in Figure 2.1. Secondary flows in axial turbomachinery appear mainly due to the viscous boundary layer of the fluid although this generally applies to any total pressure profile that enters a blade row. In a linear cascade, the radial velocity gradient in the boundary layer causes a vortex filament that rolls up into the horseshoe vortex when approaching the leading edge. This is different in a stator row where the end wall fluid might have a higher stagnation pressure than the main flow when changing from the rotating frame of reference into the stationary frame. The reason for the horseshoe vortex disappears.

In the cascade, the horseshoe vortex circumscribes the leading edge of the airfoil with two legs that trail in the downstream direction along pressure and suction sides. Both the freestream and the boundary layer flow are subjected to a cross-passage pressure gradient $\frac{dp}{d\beta}$ due to the blade force and normal to the streamlines, given by

$$\frac{dp}{d\beta} \approx \frac{u^2}{r} \tag{2.1}$$

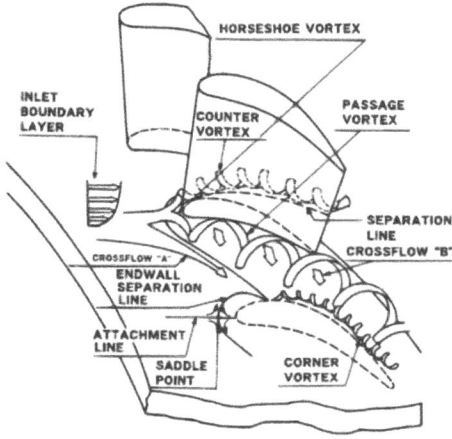

Figure 2.1: Secondary flow in a turbine taken from Takeishi et al. [91]

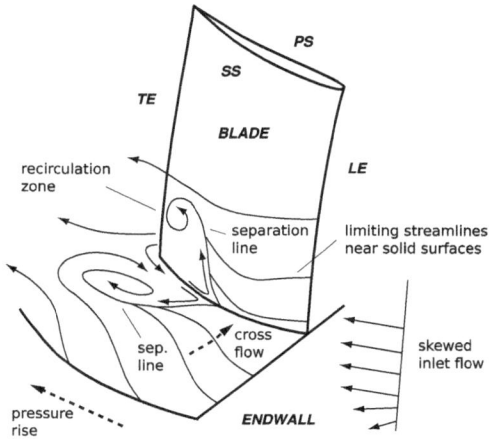

Figure 2.2: Formation of hub-corner stall with separation lines, from Lei et al. [58]

where r is the local radius of the streamline curvature. In the freestream of the blade channel, a fluid particle is kept balanced by the centrifugal force and the force of the transverse pressure gradient. The low momentum of the fluid in the boundary layer leads to a new state of equilibrium and the fluid particle has therefore to move to a smaller radius which results in a cross flow from the pressure surface side of one blade to the suction surface side of the adjacent one. The flow is overturned in the end wall regions and the strength of this flow is therefore a function of the pressure gradient which itself depends on the level of turning in the blade row. At this adjacent blade, the flow is deflected in the spanwise direction and a vortical motion is set up. In this context, the pressure surface side leg of the horseshoe vortex becomes the core of the passage vortex. Thus, the passage vortex can be seen as an additional rotational motion of the flow in a blade passage, rotating about the ideal (potential) flow direction and arising from the deflection of the inlet boundary layer. Besides that, classical secondary flow is also a function of the inlet relative total pressure profile and the aspect ratio. In comparison, a blade with a low aspect ratio will generate larger secondary flows due to the longer chord length and because a higher portion of the blade channel in spanwise direction is covered by vortices. To summarize, there are two factors that have the strongest influence on the secondary flow development. These are the formation of the horseshoe vortex at the leading edge and the strength of the transverse pressure gradient within the blade passage.

A more formal, mathematical approach was proposed by Hawthorne in his classical model of secondary flow [42]. His approach is to state that there is vorticity in the inlet boundary layer. The axis of its rotation is normal to the main stream flow and therefore called normal vorticity. This is turned as it proceeds through the downstream blade row. As a result, some part of this normal vorticity is resolved into the streamwise direction which is because the end wall boundary layer is distorted unevenly across the pitch of the blade row. As a result, the passage vortex is seen at the exit of the blade row. The shear angle between secondary and main flow leads to an increased dissipation of energy which results in pressure losses. In this context, secondary kinetic energy (SKE) describes a parameter often used to quantify secondary losses in turbines. Unfortunately, the end wall regions are the least well understood parts of a compressor and the secondary flow loss mechanism is so complex that generally valid prediction methods are not available (Denton [12]).

There are a number of important differences between compressor and turbine applications (Cumpsty [9]). At first, the flow turning in a compressor blade row is far less than in a turbine - typically 30°-40° compared to around 100° in a turbine. In turbines, the acceleration of the flow will stretch the vortices, once they have rolled up from the secondary flows, and feeding in more rotating kinetic energy. Therefore, anything that delays the initial development of secondary flows in the end wall region is expected to have an intensifying benefit [38]. Such vortex stretching will not occur in a compressor due to the deceleration of the flow. The diffusion will rather make the vortices to mix out more rapidly which can also be assumed to be the cause why smaller vortices observed in turbine rows can not be identified in compressor rows. Moreover, the inlet boundary layer of a compressor rotor is skewed in the same direction as the traverse pressure gradient, inducing a positive incidence on the blade row, unlike in turbine rotors. Both the lower turning and the fact that the incoming boundary layer acts to oppose the secondary flow driven by the pressure gradient creates the expectation of lower secondary flow losses in compressor configurations.

Another phenomenon concerning the compressor analyzed in this paper can be summarized under *three-dimensional separation*. The basic cause for the onset of separation in the region 'hub end wall - airfoil suction side' is the interaction between stagnation of low momentum fluid and blade loading. At the suction side, the low momentum fluid, being driven from the pressure to the suction side by the transverse pressure gradient, meets the low-energy boundary layer of the airfoil leading to an accumulation of low stagnation pressure fluid near the suction surface hub corner. The low momentum fluid has to overcome a streamwise increasing static pressure and will then either separate off the airfoil suction surface but still having a forward motion or develop even a reverse flow if the blade loading exceeds a critical value. If reverse flow occurs both on the airfoil suction side and the hub end wall it is referred to as *hub-corner stall*. Figure 2.2 illustrates the corresponding flow patterns. Although the consequences of 3D separations have been addressed by several authors, such as Lei et al. [58] who gives a detailed characterization of the phenomenon hub-corner stall, the nature of 3D separation and the factors that are responsible for their growth and size are not clearly understood yet. To gain a deeper understanding, Gbadebo et al. [24] presented an approach where they discussed the resulting flow pattern of 3D separations in terms of topological rules including limiting streamlines, singularities, focus and saddle points.

Harvey [38] expects the potential for reducing secondary losses in a well designed compressor row (3D blading, no corner stall at design conditions) to be less than for a typical turbine where secondary flows are much more pronounced and sees the first opportunities for a successful operation of profiled end walls in compressors retrospective applications in already existing engines. However, hub-corner stall was observed in experiments by Barankiewitz and Hathaway [1] in a four-stage compressor with 3D blading at high loading operating conditions despite this compressor having design features such as sweep and dihedral to reduce end wall loss. Therefore, non-axisymmetric end walls might have a potential benefit even in a well designed 3D compressor when blade loading is increased from design to near stall.

2.1.1 Definitions of SKE

Secondary kinetic energy (SKE) describes a parameter often used to quantify secondary losses. Basically, SKE can be seen as the energy of the flow normal to an ideal reference flow whereas the definition of the reference flow is a very challenging task. Identifying and quantifying secondary flow can therefore be ambiguous. In a linear cascade with prismatic airfoils it is quite straight forward to define the primary flow from the flow direction at mid-span (Marchal [62]). For rotating machines, cascades with 3D airfoils and non-uniform radial velocity profiles (due to an upstream rotor) this turns out to be much more complex. In the following subsections, the two definitions of SKE used during the study are presented. For the optimizations a definition was used, based on deviations from the circumferential averaged velocity components at each radial position, whereas an alternative definition based on a slip condition applied to the end wall surfaces was taken for a comparison since this definition is trusted to capture the secondary flow structures more accurately. However, the second definition was not considered for the stator optimization due the enormous separations of the original stator design which would not have allowed any reasonable comparison of the 'classical' secondary flow phenomena between the datum and the optimized stator geometries.

SKE Method 1 - Pitchwise Averaged Velocity Profiles

The first definition is based on deviations from the pitchwise averaged velocity components at each radial position. This method is quite common and was used by Brennan et al. [7], amongst others, who additionally weighted the SKE by the helicity to filter the potential effect of the airfoil. As there is no general definition of secondary flows available in the NUMECA environment (which is quite comprehensible as there no universally applicable definition that holds true for any kind of turbomachinery issue) the following procedure had to be externally implemented by scripting:

At a z-constant plane (which the user has to define) the cylindrical velocity components are pitchwise averaged along each radial grid line, resulting in radial velocity profiles $V_r(r)$, $V_\varphi(r)$ and $V_z(r)$ for the corresponding evaluation plane. The primary flow direction is then defined by normalizing this vector of the radial velocity profiles, leading to the unit vector

$$\vec{V}_{prim}(r) = \frac{1}{\sqrt{(V_r(r)^2 + V_\varphi(r)^2 + V_z(r)^2)}} \begin{bmatrix} V_r(r) \\ V_\varphi(r) \\ V_z(r) \end{bmatrix}. \tag{2.2}$$

To define the velocity component (scalar) in the primary flow direction of every grid point, the local velocity vector (i.e. depending on r and φ) is projected via dot product onto the primary flow direction for the corresponding radial position. A multiplication by the unit vector of the primary flow direction leads to the local primary flow vector.

$$\vec{V}_{prim}(r, \varphi) = \left[\vec{V}_{prim}(r) \cdot \vec{V}(r, \varphi) \right] \vec{V}_{prim}(r) \tag{2.3}$$

The secondary velocity is the local deviation from the primary velocity.

$$\vec{V}_{sec}(r, \varphi) = \vec{V}(r, \varphi) - \vec{V}_{prim}(r, \varphi) \tag{2.4}$$

The secondary kinetic energy is now defined as

$$SKE = \frac{1}{2}|\vec{V}_{sec}(r,\varphi)|^2 \tag{2.5}$$

and although the density is missing in the derived expression, which makes referring this term to kinetic energy misleading to some extent, it will be compensated later on in the optimization process by using the mass-weighted averaged SKE values at the evaluation planes defined by the user.

Since this definition can be derived directly from the CFD calculation it was used for the optimizations. Yet, this method has the disadvantage that the secondary flow is not known for the entire flow domain. The user has to define all z-constant sections of interest which may make post-processing quite time-consuming depending on the number of evaluation planes. Moreover, this definition will be influenced by the potential field of the blade row. Within the blade row its effect on the flow field will appear as SKE which will be spurious. The potential effect could be excluded by weighting the SKE by the helicity (Brennan et al. [7]). Helicity describes the flux of streamwise vorticity but, nonetheless, does still not exclude the potential effect in areas of vortical flow. Since only the SKE at the stator exit plane was evaluated throughout the optimization process, the error introduced by the potential field was expected to be small.

SKE Method 2 - Euler Wall

The second definition is based on the slip condition applied to the hub and casing end walls, eliminating the boundary layer. The principle of this idea has already been used by Niehuis et al. [69], who used an inviscid Euler computation to define the primary flow velocities for evaluating the SKE. However, an inviscid Euler computation can normally not be run on the same mesh as the Navier-Stokes computation, as the solution tends to diverge. To obtain a convergent solution, the mesh has therefore to be coarsened which, in turn, requires an interpolation for the post-processing to define the corresponding primary flow for each cell of the Navier-Stokes computation. A combined approach was introduced to overcome this drawback of the pure Euler computation. It contains a Navier-Stokes simulation with the same grid as the datum simulation, the end walls being defined as inviscid Euler walls. Then, on condition that an one-dimensional incoming flow in axial direction is assumed, the vortex filament resulting from

$$rotV_\varphi = -\frac{\partial V_z}{\partial r} = 0 \tag{2.6}$$

is eliminated and, hence, also the source for the horseshoe vortex. No unbalance between centrifugal force and the static pressure force would occur in the end wall region due to the absence of the boundary layer with low momentum fluid. Thus, no cross flow would be visible in the blade channel.

In the author's opinion, this definition is far better, since the main effects of secondary flow in the blade channel can be isolated. The subtraction of the Euler wall computation from the original Navier-Stokes computation should then enable visualization of the secondary flow structures. The evaluation of the secondary flow is similar to the approach that uses pitchwise averaged velocity profiles. The main difference is that the SKE can be evaluated for each grid cell, since identical meshes are used for both simulations.

$$\vec{V}_{prim\_NS}(r,\varphi,z) = \left[\frac{\vec{V}_{EW}(r,\varphi,z)}{|\vec{V}_{EW}(r,\varphi,z)|} \cdot \vec{V}_{NS}(r,\varphi,z) \right] \frac{\vec{V}_{EW}(r,\varphi,z)}{|\vec{V}_{EW}(r,\varphi,z)|} \tag{2.7}$$

now defines the reference flow vector for each cell and

$$\vec{V}_{sec\_NS}(r,\varphi,z) = \vec{V}_{NS}(r,\varphi,z) - \vec{V}_{prim\_NS}(r,\varphi,z) \tag{2.8}$$

the secondary flow vector.

The big advantage of this method is that the secondary flow is known in the entire flow domain and arbitrary cutting planes can be evaluated. Unfortunately, this method always requires a second flow computation which makes it rather expensive in terms of CPU-costs. Therefore, it was not embedded within the optimization process. Only for the original axisymmetric geometry and the different optimized designs additional simulations with Euler walls were carried out and compared to the result from the first SKE definition. Moreover, a lack of velocity can be observed outside of the boundary layer when this simulation is compared to a simulation with viscid walls both run at the identical operating point with the same mass flow. Then, of course, mass flow is shifted from the mid-sections towards the end wall regions.

2.2 Methods to Control End Wall Flow

Before dealing with the different methods to control end wall flow in more detail, a few general design considerations shall be given. In the design of any compressor the initial decisions on the layout and duty mainly determine the problems to be encountered and the level of efficiency to be achieved. Today, stage loading, which is usually expressed by the pressure rise in relation to the stage numbers and the rotational speed is recognized to be the single most important design decision. After the calculations performed at a mean radius in the preliminary design, refinements are introduced to assess the blade loadings at hub and casing. A criteria for satisfactory blade loading that is still considered within any compressor design process can be described by the diffusion factor or alternatively the equivalent diffusion ratio, both traced back to Lieblein et al. [60]. They stated that the total pressure loss coefficient of the blade was connected to the momentum thickness of the wake

$$\omega = 2\left(\frac{\theta}{c}\right) \frac{\sigma}{\cos\alpha_2} \left[\frac{\cos\alpha_1}{\cos\alpha_2}\right]^2 \qquad \text{with} \qquad \sigma = \frac{c}{s}. \tag{2.9}$$

They recognized the boundary layer as the critical factor for the turning and the pressure rise of a blade which would be mainly affected by the deceleration of the flow. To capture this aspect, Lieblein defined a local diffusion factor which basically relates the peak velocity on the suction surface of the blade to the velocity at the trailing edge. Since it was quite complex to evaluate this factor in the past due to the need to assess the blade surface pressure distribution, Lieblein derived a simplified diffusion factor which is given by

$$D = 1 - \frac{c_2}{c_1} + \frac{\Delta c_u}{2\sigma c_1} \tag{2.10}$$

for a two dimensional blade section. The first component is due to the one-dimensional deceleration of the flow with c_1 and c_2 as the corresponding velocities into and out of the blade row. The second term describes the turning of the flow with Δc_u as the change in whirl velocity. This second term also introduces the solidity expressed by the chord to blade pitch ratio. For a rotating row, the velocities have to be replaced by the relative velocities. In summary, the diffusion factor describes the boundary layer load where values greater than 0.5 are considered as critical and values that exceed 0.6 indicate blade stall and the limit for confident operation of blades. Values around 0.45 can be taken as a general design choice. Lieblein later defined an alternative approach for looking at blade performance introducing a diffusion ratio a more detailed description of which can be found in Cumpsty [9].

Often, it is not sufficient to focus on profile losses alone (especially for the evaluation of high-lift design concepts) but rather should one pay attention on minimizing the total row loss including the contribution of the end wall regions. Unfortunately, it has not been possible to completely adopt the criterion of Lieblein for end wall loading yet because the fluid mechanics in these regions are still not entirely understood. In this field, a very promising approach is the criterion proposed by Lei et al. [58] for estimating the size and strength of three-dimensional hub-corner stall in rotors and shrouded stators of multi-stage axial compressors. The idea behind this approach is the introduction of a diffusion factor which defines at what conditions hub-corner stall will occur and a stall indicator S which is expressed by the difference of the Zweifel blade loading coefficient at mid-span and in the end wall region. This stall indicator defines whether flow reversal exists on the blade suction surface and the hub end wall. Nonetheless, this topic is still a matter of research and there is no general method established, yet, that quantifies end wall load. This is of crucial importance as the profile losses are of small order around $\omega < 0.02$ and contribute only a small fraction of the total losses in a typical blade row operating near its optimum [9]. Of more importance than the blade losses is that the blades should produce the designed deflection enabling the flow to leave the blade row at the desired outflow angle. This, again, is particularly relevant in the end wall regions. In the following subchapters, three methods are described which either influence blade loading in the end wall region or at least manipulates end wall flow.

2.2.1 Non-Axisymmetric End Wall Profiling

The application of non-axisymmetric end walls is one way to reduce blade loading in the end wall regions and to control end wall flow when the main objectives are to increase component efficiency and total pressure ratio. Unlike traditional design, where hub and casing are surfaces of revolution, non-axisymmetric profiling leads to fully three-dimensional hub and casing surfaces which has different effects on the flow field. End wall profiling works according to the principle of streamline curvature and is illustrated in Figure 2.3. Convexity will locally drop the static pressure and accelerate the flow close to the end wall, concavity will do the reverse due to the local diffusion. This effect is local and occurs only in the end wall region. A second mechanism is the change of the cross sections which can, depending on the aspect ratio, affect the complete flow field over the span, a third effect is the modification of blockage through thickening of the boundary layer. The objective is to diminish the traverse pressure gradient as the driving force behind secondary flow by raising and decreasing the static pressure on the suction and pressure surfaces respectively. This does not really lead to a global unloading of the

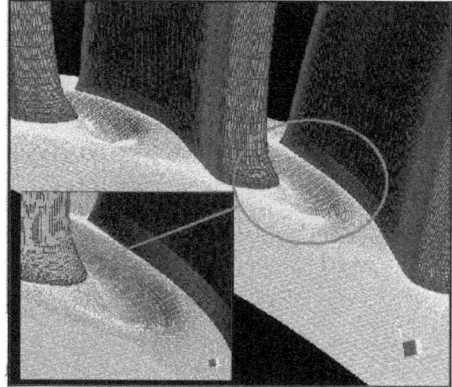

$$p_{stat} \uparrow, \vec{c} \downarrow$$

Concave curvature

$$p_{stat} \downarrow, \vec{c} \uparrow$$

Convex curvature

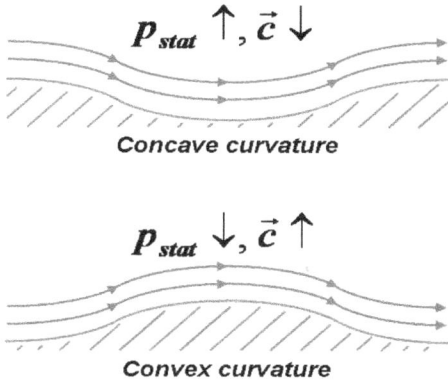

Figure 2.3: Influencing static pressure and flow velocity by end wall profiling

Figure 2.4: Example for the application of non-axisymmetric end walls in a turbine cascade [70]

profile but rather to an aft-loading of the blade. Even though aft-loading is generally considered to lessen secondary flow and loss, one should act with caution concerning aft-loaded airfoils to avoid overloading. Although this methodology initially intends to modify the end wall flow most of the design methods may also have consequences on the whole flow field due to the secondly mentioned mechanism and, which counts particularly in low aspect ratio passages, the dominating influence of the walls.

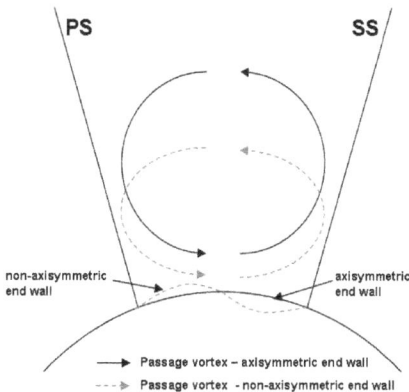

PS   SS

non-axisymmetric end wall

axisymmetric end wall

→ Passage vortex – axisymmetric end wall
- - ▶ Passage vortex - non-axisymmetric end wall

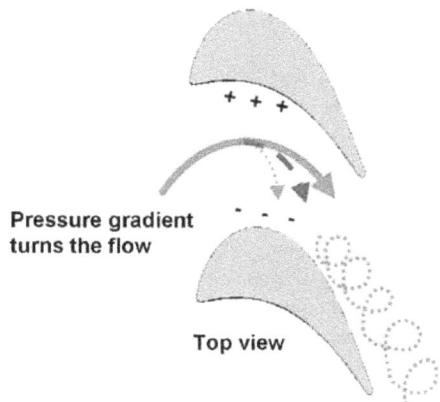

Pressure gradient turns the flow

+ + +

- - -

Top view

Figure 2.5: Effect of end wall profiling on the passage vortex and exit flow field

Figure 2.6: Overturning of the end wall boundary layer in a blade passage [37]

One aspect also to be mentioned is the classic effect of end wall profiling on the exit flow field. The part of the passage vortex nearest the end wall will cause overturning of the flow while that part of the vortex most inboard of the wall will cause underturning of the flow. Although the intended purpose of non-axisymmetric end walls is to reduce the cross passage pressure gradient and therewith cross flow, surprisingly, a higher overturning of the boundary layer is frequently observed when profiled end walls are applied (Brennan et al. [7], Torre et al. [92], Germain et al. [26]). These publications on profiled end walls report that the loss cores happen to be confined closer to the end walls compared to the planar references as schematically illustrated in Figure 2.5. Once an inlet boundary layer or any rotary stagnation pressure gradient has been turned through a blade row there must be some streamwise vorticity in the exit flow field. Whatever 3D shaping is applied to the airfoil and/or its end walls this vorticity cannot be eliminated. What can be done is to ideally minimize the energy contained in the vortex and its impact on the aerodynamics of any downstream row. This may be achieved by delaying the formation of the passage vortex which provides a second explanation for the higher overturning. In the ideal case, the end wall flow would not impact on the suction side of the adjacent blade and would not have the time to roll up into a vortex as indicated by the comparison between the dark and light blue arrows in Figure 2.6. This would then be seen as a highly overturned end wall boundary layer although the cross passage pressure gradient might have been reduced. Although addressed in more detail in the next chapter, some publications shall already be mentioned in which this behavior was observed. In Torre et al.[92] the passage vortex did not reach the suction side and did not interact with the trailing edge vortex. As a consequence, the passage vortex remained closer to the end wall and as a result the part of the passage vortex which causes overturning. Brennan et al. [7] also reported on a higher overturning near the end wall leading to increased incidence angles on the roots of the the downstream row. He similarly reasoned that some of the low momentum fluid that used to sweep into the passage vortex for the axisymmetric design remained on the end wall and led to higher end wall losses.

Turbine Applications

Full passage end wall contouring for loss reduction has been studied by a large number of researchers in the past 15 years and is now well established and successfully applied to axial flow turbines, see Figure 2.4 as an example. The successful application of profiled end wall begam with the CFD study of Rose [79] whose attempt was to reduce the circumferential non-uniformities in the static pressure at the trailing edge platform between a HP NGV and the corresponding HP rotor by re-shaping the hub end wall. Streamline curvature was employed in the radial direction in order to counteract accelerations in the tangential direction. A sinusoidal profile was applied in the circumferential direction which equalized the static pressure and re-duced the peak pressure at the exit of the NGV with the ultimate objective to minimize purge flow requirements. Although the predictions were later successfully confirmed in the experi-ments by Hartland [35] Rose cautions that end wall contouring causes local changes in airfoil loading and could potentially outweight any benefit of the 3D shape.

Hartland et al. [34] confirmed the possibility of controlling the end wall static pressure using profiled end walls in the Durham linear cascade. The design proposed by Rose [79] was there-fore slightly modified for the rotor blade. A significant reduction in the strength of the cross passage pressure gradient early in the passage was achieved by relieving the front part of the blade which yielded a drop of the mixed-out secondary loss by 34%, a reduction of SKEH and a smaller angle deviation at cascade exit. The decreased overturning close to the end wall was found to be due to a countervortex rotating in the opposite direction and beneath the passage vortex. The countervortex in turn resulted from an enhancement of the suction side horseshoe vortex due to a ridge located upstream of the leading edge. This work was followed up by Ingram et al. [54] who described detailed measurements of the aforementioned design in the Durham cascade introducing improved measurement techniques with better capabilities to capture the flow near the end wall and compared the results to a second, more sophisticated, profiled end wall design that was restricted to the blade passage. This second design did not show the upstream and downstream ridges which is why this design had been expected to further reduce losses while accepting some penalty in terms of overturning. However, the observed loss reduction had been smaller than expected in previous experimental investigations [29]. With the help of the new measurement equipment, these contradictions between the first and second end wall designs could be clarified. The new testing demonstrated only a 18% reduction in mixed-out secondary loss because of the better resolution of the corner vortex while the second design showed, as assumed, a higher decrease by 24% due to avoiding the undesired counterrotating vortex although this led to higher overturning. Moreover, the strong effect of end wall contouring on the flow pattern in the early blade row was emphasized which then translated into a reduced loss in the aft part of the row.

Brennan [7] describes the redesign of the single stage HP turbine (nozzle and rotor) of the Rolls-Royce Trent 500 engine. A new linear design system for the parametric definition of the end walls was used consisting of a predetermined number of discrete axial stations along the mean airfoil camber line. In this design system, a sinusoidal shape is generated at each axial position in circumferential direction, the height and phase angle of which have to be defined by the user. These perturbations are described by a harmonic series of zero- and first-order where a zero-order harmonic implies an axisymmetric change of the annulus geometry. The design goal was to minimize SKEH and whirl angle deviation due to the Code's drawback to predict mixing processes accurately enough and therefore to capture the total row loss. Certainly, the poor loss prediction was also related to the coarse mesh compared to today's standards. Based on the experimental work of Hartland et al. [34], the assumed proportionality between SKEH and secondary loss was used in order to estimate the entire improvement in stage efficiency to be 0.4% (0.24% allotted to the NGV and 0.16% to the rotor). This prediction was exceeded in the test rig validation by Rose et al. [80] who reported a 0.59% improvement in the stage efficiency. However, relying on the same correlation between SKE and end wall loss, Ingram et al. [54] published cascade test results where the provided SKE reduction did not lead to any decrease in pressure loss. This emphasizes again the difficulty to fully understand the source of end wall loss.

The research on the Rolls-Royce Trent 500 engine was continued by Harvey et al. [40] who provided a redesign of the engine's IP turbine in a multi-row environment based on the previously mentioned work, i.e. the HP turbine having profiled end walls in the datum arrangement. The optimization yielded a SKEH reduction of 38% for the turbine vane and 28% for the rotor corresponding to 0.96% decrease in mixed out loss according to the correlation of Hartland et al. [34] (again, it was not possible to capture the changes in loss accurately enough in the CFD calculations). This range was then experimentally confirmed by cold flow rig testing. Two more important aspects were addressed in this work. In previous publications, end wall contouring had been shown to push secondary flow closer to the end wall worsening the hub and casing inflow conditions in the downstream blade due to higher incidences. Yet, this negative impact could not be confirmed in this multi-row investigation. Secondly, the sensitivity of the effect of profiled end walls to design speed, Mach number and the corresponding incidences was documented. The benefits of the contoured design were best below nominal design values which had been some kind of unexpected since the investigations on the HP turbine alone by Rose et al. [80] had shown a different behavior (e.g. at 120% rotational speed the redesign of the HP turbine did not offer any improvement in stage efficiency while it remarkably did so for the IP turbine). The authors concluded that off-design performance including the sensitivity to positive incidence and exit Mach number must be considered more strongly during the design.

At the same time, Eymann et al. [18] addressed airfoil thickening in combination with axisymmetric end wall profiling in a three-stage LP turbine whereas only the first stage was modified and tested with respect to the development of secondary flow in the entire arrangement. Although a reduction of secondary loss could be achieved this was partially outweighted by increased profile losses. Mainly, this additional loss was created inside the wake of the altered vane.

Nagel and Baier applied a parametrization to both the airfoils and the surfaces of the annulus of the T106D cascade [67] to make full use of the available geometrical degrees of freedom and employed a gradient-based optimization process to find the best possible design. In contrast to most of the aforementioned studies, the total pressure loss predicted by the CFD simulations was taken directly as the key design objective in this work which was possible due to a comparatively fine grid. The corresponding experimental work validated the significant decrease in row loss by 22% which had been predicted by the CFD. The position of the loss cores and their values proved that modern CFD codes with fine grids were capable to capture loss mechanisms at least to some extent. Furthermore, SKE was reduced by 60%.

A further work of Ingram et al. [53] reports on a previously unreported secondary flow feature for turbine blading. They investigated a rather ambitious design with very high peak-to-dip amplitudes in order to determine the application limits of profiled end walls. This design locally separated the inlet boundary layer which was accompanied by a strong vertical flow up the suction surface towards mid-span. As expected from former research, the loss core of the PEW design was located closer to the end wall but it was not considerably smaller than for the planar cascade. Significant extra loss was generated despite the secondary kinetic energy was largely reduced. Obviously, the high amplitudes of the profiling resulted in an excessive increase of the static pressure at the suction side involving this new radial flow driven by the increased spanwise static pressure gradient. However, the CFD was not able to reproduce this phenomenon. The authors assumed that this was due to the limitation of the used algebraic turbulence model.

Torre et al. [92] presented an approach for reduction of secondary flow within the blade passage by modifying the end wall geometry of a cascade with solid-thin profiles typical for LP turbines. In contrast to most of the previous studies, which had considered the transverse pressure gradient to be the key flow parameter to affect secondary flow, this work was focused on the formation and development of the horseshoe vortex. Other techniques, such as leading edge modification, had already shown the potential to affect the onset of secondary flow. Sauer et al. [82], for instance, placed a leading edge bulb to control the formation of the horseshoe vortex. However, the authors expected end wall profiling as a more powerful tool as the entire flow field along the whole passage could be affected. The idea of Torre's approach is based on the fact that a reduction of the front loading leads to a weakening of the pressure side leg of the horseshoe vortex by entraining a greater amount of the inlet boundary layer into the suction side leg. That way, vorticity is transferred to the suction side leg and the development of the pressure side leg and the consequential passage vortex is delayed because the suction side leg keeps it away from the blade suction surface. This design approach led to a radial depression up to 20% of the blade pitch concentrated near the leading edge. A decrease of 72% in SKEH and 20% in mixed out end wall loss could be measured in the experiments. Moreover, a high level of agreement with the experimental data could be revealed by the CFD results both in reproducing the effect of the end wall application and the reduction of SKEH and losses.

Besides the investigation of secondary flow features, also a lot of effort has been put into the end wall design systems to control these features. In 2007, Nguyen and Squires [68] presented a simple method to deform the hub end wall of a turbine nozzle blade row with a single geometry function which requires two parametres determined by a CFD-based optimization. For this approach, the local change in radius is directly proportional to the local tangential pressure gradient. They used the discrete approach with several axial stations in order to modify the hub end wall. Unlike other authors, who used two separate functions (one in the axial, one in the tangential direction), their objective was to use only one single transformation function which would depend on the radius change amplitude $K(z)$ for the manipulation in axial direction and a phase shift angle ϕ for the manipulation in circumferential direction. $K(z)$ in turn, would be proportional to the local pressure gradient $K(z) = \alpha \Delta P(z)$. A parameter study was carried out analyzing a combination of 6 values of the phase shift angle and four values of the proportionality coefficient α respectively. The authors report on a reduction in SKE at the blade trailing edge by approximately 17% at design conditions for the best sample. However, the authors do not comment on whether the optimum values for the parameters have been found or not. A further advancement for this promising approach could for example be, to exploit a genetic algorithm for optimization purpose to find the best suitable values for these parameters.

In the same year, Praisner et al. [73] published a study on the application of non-axisymmetric end wall contouring to a variety of three turbine airfoils, comprising one conventional and two high-lift LP turbine airfoil designs, to minimize the end wall losses. Both high-lift airfoils featured a 25% increase in loading compared to conventional designs, one being aft- and one being front-loaded. A gradient-based optimization algorithm was coupled with a CFD solver in order to vary a free-form parametrization of the end wall surface. Since the CFD solver had already successfully demonstrated to predict loss modifications in the Langston cascade, the mass-averaged magnitude of viscous losses was employed as the key parameter for the optimization methodology. The authors found the greatest benefit for the front-loaded high-lift airfoil and could demonstrate for this design the potential of profiled end walls to reduce losses primarily associated with the passage vortex. This is in keeping with previously published evidence that front-loaded airfoils show higher secondary flows and loss compared to aft-loaded designs [95]. Consequently, this end wall design also showed the most pronounced topography compared to the other designs. The predicted reduction in row loss for the front-loaded high-lift airfoil by CFD was 12% while the measured loss decrease was at 25% even twice as high, revealing once again that even state-of-the-art CFD solver still have a lack of predicting losses accurately enough especially when it comes to individual flow features such as the passage vortex.

In order to improve the understanding of the secondary flow features, Germain et al. [25] presented a new visualization criterion with the aim to isolate the vortical flow structures from the complexity of the three-dimensional flows that are responsible for loss generation. According to the authors the overall complex multi-row interaction should be taken into account when trying to increase efficiency in a multi-stage environment and the goal of these attempts should not be limited to the reduction of loss in each stage. Therefore, the objective of this study was to derive more powerful post-processing tools to enable a better interpretation of numerical results concerning the complexity of end wall flows. A modified definition of the SKE was provided which the authors assumed to be more suitable to turbomachinery applications. Their main concern regarding the definition based on pitchwise averaged velocities for each radial position was that it did not account for radial non-homogeneities, such as over- and underturning. Their alternative definition filters out the flow inhomogeneities both in pitchwise and spanwise direction by using certain radial windows at each pitchwise position for the radial averaging of the velocities which allows to capture over- and underturning. Additionally, a flow field topology based vortex criterion was introduced. Vorticity alone, although fundamental for the understanding of the secondary flow physics, is not sufficient to isolate the different flow features since boundary layers, wakes and vortices are characterized by vorticity. To identify the vortical structures, the authors resolved the eigenvalues of the velocity gradient tensor. In this context, complex eigenvalues are related to vortices and the vortex intensity is represented by the magnitude of the imaginary part of the eigenvalues [81]. With the vortex strength criterion, the authors were able to distinguish very clearly between passage, corner and trailing edge vortices. For validation purpose, the end walls of a large scale linear cascade were then redesigned employing these new parameters. Two optimizations relying on different sets of parameters were conducted and afterwards experimentally checked. The resulting designs showed quite distinct mechanisms for loss reduction. For the first design, the power of the transverse pressure gradient in the blade channel could be reduced leading to a large reduction in SKE. In contrast, the second design showed changes only in the early part of the blade passage. Nonetheless, both designs led to a similar reduction in secondary loss and for the second design, an enhancement of the suction side leg of the horseshoe vortex could be visualized by the vortex strength criterion keeping the passage vortex away from the suction side - i.e the same mechanism that had already been reported by other researches such as Torre et al. [92]. This depicts again the different mechanisms, i.e. either the formation of the horseshoe vortex or the transverse pressure gradient, that are linked to the possibility to reduce secondary loss.

In a follow-up study Germain et al. [26] applied non-axisymmetric end walls to a one-and-a-half stage high work axial flow turbine using a combination of SKE and total pressure loss as a design target. For the end wall design, the effect of the fillet radii was additionally considered. The authors report a stage efficiency improvement of 1% in the related experimental testing. The main reason for the improvement was found to rely on the loss reduction in the first row and the more uniform flow field. Especially for the vane, additional improvement for the mid-span flow could be observed due to the low aspect ratio characteristic. In the CFD, the loss reduction was considerably underestimated in the tip region predicting a rather small overall benefit of 0.2%. Moreover, the CFD was not able to predict the weakening of the profile losses over the entire core region. The numerical investigations indicated for the selected design that especially the settings of the transition model used for the airfoil had a major influence on the loss predictions since the secondary flow structures were obviously triggered by strong transitional effects. This is particularly crucial if the end wall boundary layer is also of transitional nature which had not been modeled for the test case. This issue may have less impact for higher Reynolds numbers which is the case e.g. for transonic compressors where it is normally sufficient to model the boundary layers as fully turbulent. In the second part of the study, Schuepach et al. [83] provide the analysis of the corresponding time-resolved flow physics in order describe the discrepancies in more detail. By evaluation of the time-dependent pitch angle variation at constant spanwise positions at the IGV outlet a weaker circumferential pitch angle gradient could be detected for the profiled geometry. This weaker gradient in turn indicates a weaker trailing edge vorticity. Therefore, less trailing shed vorticity and less dissipation was found inside the wake for the profiled case resulting in the observed loss reduction in the mid-span region. This effect mainly results from the interaction with the upstream potential field of the rotor. Hence, it is not surprising that this effect could not be captured in the steady CFD simulations where such information are filtered out by the mixing planes at the rotor-stator-interfaces and not further transmitted into the corresponding upstream and downstream blade rows respectively. In that case, unsteady CFD simulation will most likely help to improve the numerical predictions but of course at the expense of a much higher numerical cost. However, this corroborates to some extent previous publications concerning the still existing problem of CFD that accurate loss predictions is even today a matter of concern.

Sonoda et al. [87] used a stochastic optimization method to analyze the effect of axisymmetric end wall contouring on the performance of the inlet guide vane of an ultra-low aspect ratio turbine to match the requirements of small turbofan engines for compact business jets. Due to the very low aspect ratio of 0.21 the influence of secondary flow on turbine efficiency was expected to be very high. Since the flow was supersonic at a Reynolds number of $3.5x10^6$ the authors also intended to investigate differences between the occurring secondary flow and the modell derived from low speed, linear cascade conditions. Both independent hub and tip modifications and combined end wall modifications were employed to observe the correlations between the changes and their influences on the pressure loss. To find the optimum shapes a single objective function was minimized by the optimization method which included pressure loss and constraints on outflow angle and mass flow rate within certain ranges. For all optimized cases the passage diverged around the frontal part and converged around the rear part of the blade passage. This involved an aft-loading of all profiles which is known to produce less secondary flow. Within that study, a maximum loss reduction of 10 % compared to the datum design could be achieved for the combination of hub and tip contouring. Concerning the secondary flow model, a tendency of underturning was observed near the hub. The authors concluded that this tendency was predominant in case the exit flow was supersonic but could not provide a clear explanation for this behavior.

Apart from the already described aspects, Mahmood et al. [61] and Gustafson et al. [32] additionally showed the influence of non-axisymmetric end walls on the passage heat transfer. Secondary flows increase the thermal load on a turbine passage wall by lifting up the cooling flow which leads to a reduced film cooling efficiency and therewith to higher cooling requirements. Mainly, the passage vortex and the suction side leg of the horseshoe vortex cause the film cooling jets to lift off from the end wall which could be documented in film cooling measurements by Friedrichs et al. [21]. To investigate this phenomenon further, Mahmood et al. [61] made a comparison between 3D profiled end walls and leading edge fillets where the fillets were applied to the corner of the leading edge extending about one third of the axial chord length. In order to deduce heat transfer, the Nusselt numbers on the end wall were measured. The static pressure measurements did not show any effect of the fillets on the blade surface but a mitigated pressure gradient on the end wall. For both analyzed test cases, smaller end wall Nusselt numbers and corresponding smaller rates in heat transfer were observed especially in the upstream part of the throat region. However, end wall Nusselt numbers and heat transfer were reduced to a greater extend by the leading edge fillets while the pressure losses were found to be lower with the contoured end walls. Gustafson et al. [32] addressed how secondary flow interfered with the end wall cooling jets. Yet, the main goal of their study was to provide detailed measurements of the in-passage velocity and pressure fields. The authors stated that earlier studies had mostly been limited to data at the passage exit plane and they hoped to gain a better understanding and a more complete picture of the evolution of the vortex system inside a cascade with non-axisymmetric end walls. For the measurements, realistic Reynolds numbers were employed while the Mach numbers were well below engine conditions. This was accepted as it was intended to derive qualitative guidelines from the study. Concerning heat transfer the onset of the horseshoe vortex was moved further upstream of the stagnation point compared to the axisymmetric baseline design. The region around the stagnation point is normally an area of high heat transfer due to the presence of the horseshoe vortex. Therefore, also the heat transfer could be reduced in this region by displacing the horseshoe vortex. Furthermore, the turbulence intensity were measured showing reduced intensities near the end wall for the contoured case which is supposed to decrease heat transfer on the end wall due to the lower momentum transfer.

An example for the successful retrospective application of non-axisymmetric end walls in an industrial gas turbine is given by Blackburn et al. [5]. Incidentally, this is the only one known to the author from literature. They report on a redesign of the industrial Avon gas turbine which had been derived from the first Rolls-Royce jet engine to enter production which contained an axial compressor. The redesign was focused on the three-stage turbine including, apart from the target to deliver power and efficiency gains, additional key requirements, such as compability with multiple engine standards and mechanical attributes. The airfoil design was changed by aft-loading the profile which resulted in an acceleration through the throat plane. This enabled to control diffusion and to reduce secondary loss in the early part of the passage. Thereafter, profiled end walls were applied to better control the migration of vortical structures. Special attention was paid to avoid separation that may have occurred due to the excessive local diffusion as a consequence of the higher acceleration. This led to a decrease of 30% in secondary loss. With all taken actions, an overall increase of 2.5% in LP turbine stage efficiency could be achieved where the improvements of the retrospectively applied non-axisymmetric end wall were predicted to contribute a fraction of 0.4%.

Compressor Applications

All the above mentioned work is related to axial flow turbines. Not until recently have researchers started to investigate the potential of non-axisymmetric end walls for compressor applications and therefore only a few publications on this topic are available. One of the first researchers addressing this subject were Hoeger et al. [49] who investigated, even though restricted to axisymmetric design, the influence of a concave hub shape on the pre-shock Mach number and the shock pattern. Introducing this new cascade technique the profile boundary layer could be unloaded at the expense of an increased hub shock loss by changing the shock system from oblique shocks in the linear contracting cascade to a normal shock pattern in the concave end wall shape. In a further work of Hoeger et al. [50], this difficulty was solved by applying a rather simple non-axisymmetric end wall contour consisting of a combination of a axisymmetric contour on the pressure side and a linear hub shape on the suction side being connected by a sine-type function in pitchwise direction. This allowed to reduce blade loading without any rise in shock strength and losses. Dorfner et al. [16] used a new design system connected to an optimization scheme for the investigation into non-axisymmetric end walls in the 3rd row of the last stage of the RWTH Aachen three-stage research compressor IDAC3. They used a sophisticated approach for end wall modification applying B-spline tensor product surfaces for geometry manipulation while three major operating points were considered for the optimization process. In order to validate their optimization approach, a stagger-line improvement was performed as a first exercise as the expected result was known. After the validation, it could be shown in a second design exercise how the flow field could be strongly influenced by profiled end walls.

Iliopoulou et al. [52] presented a design of a non-axisymmetric hub end wall of a HP compressor rotor blade for performance improvement. For this purpose, a genetic algorithm was employed and combined with a surrogate model of radial basis functions in order to overcome the genetic algorithm's slow convergence due to its stochastic nature. In a first step, they optimized the blade shape with different tip gap values to validate the optimization chain. The subsequent design of the profiled end wall led to an 0.4% increase in stage efficiency. This gain happened rather by changing the shock mechanism than reducing the secondary flow and loss as no considerable decrease of the blade loading was observed. The obtained improvement was comparatively high since the blade operated in the transonic region across the entire span which resulted in extra losses in the whole channel. However, the total pressure ratio was decreased by 0.4% compared to the axisymmetric case at design conditions. Moreover, the optimization resulted in a lower efficiency close to stall which the authors traced back to the fact that only one point had been considered during the design process. Therefore, they suggested to take into account several operating points, one of which should be close to stall, for future work to provide a more robust optimum. Nonetheless, the question remains if this achievement could have also been achieved by only modifying the 2D axisymmetric contour like in the work of Hoeger et al. [49].

This work was pursued by the same team of authors with a design study on different end wall shapes for a booster's 2nd stage rotor, Lepot el al. [59]. In order to escape from the restriction of local optima the same optimization scheme was used considering two operating points. The design objective was to maximize stage efficiency at design conditions while preserving the total pressure ratio at stall to guarantee the same stability. Additionally, the authors intended to estimate the impact of fillets on the end wall design. Due to the limitations of the grid generator of the optimization chain (the grid generator was able to handle either fillets or non-axisymmetric end wall but not both features together in one project) the fillets were, however, not included in the optimization but analyzed in posterior simulations after a manual application by an external design system. The profiled end wall design offered a rise in efficiency of 0.15% while the axisymmetric yielded an even higher increase of 0.2%. Analyzing the flow features, the migration of the suction side corner vortex was observed to start later for both contours reducing the strength of the vortex. At the same time, its size remained constant as indicated by the radial extension and the suction side peak Mach number was smoothed out. Both in terms of perturbation and flow features the axisymmetric contouring appeared as an azimuthal average of the profiled end wall design which again raises the question to what extent the effect of the non-axisymmetry really contributes to the improvements. The posterior verification of the fillets only showed a small influence for the profiled end wall design and no impact for the original design. In both cases, the performance appeared preserved. However, this may be different if the fillets were included within the optimization procedure since the end wall design would then probably look different.

In 2008, Harvey presented a two-part paper on a linear cascade investigation [38] and a multi-stage, HPC, CFD study [39]. The experimental investigations on a linear cascade of a compressor stator at Cambridge University demonstrated improvements to the exit flow field in terms of local reduction in loss and underturning in the secondary flow region [38]. Besides that, the accompanying CFD results showed very good agreement with measurements taken at design conditions and a qualitative match at off-design. The design methodology adopted was that of one of his former works (Brennan et al. [7]) which in the meantime had been superseded by modern optimization techniques and therefore had not been expected to be optimum. It was intended to asses the potential for the control of secondary flow and corner separation and relate this to the known beneficial effects of dihedral and sweep in compressor design. For this reason, the results of the profiled end wall geometry were compared to the ones of the corresponding reference, prismatic and leant airfoils. Both the contoured end wall and the leant airfoil proved to reduce maximum under-turning by 1.5° in the end wall region and total mixed out loss by 7%. However, he observed an increased overturning in the end wall boundary layer.

End wall experiments in linear cascades have, unfortunately, little relevance to compressor blade rows [9]. Linear cascades typically have collateral inlet boundary layers of which the inlet flow direction is uniform and the same as the rest of the inlet flow. This is quite different from a real compressor where the inlet boundary is highly skewed. This cannot be replicated in a linear cascade without a complex moving end wall arrangement at the inlet. Thus, the best approach to gain a deeper understanding of the flow mechanisms for end wall losses are 3D CFD analysis of non-axisymmetric end walls undertaken at real engine conditions. This has been done in the second part of Harvey's work [39] in which he applied profiled end walls by means of an optimization in a typical embedded stator row in a 6 stage axial flow research HP compressor. This compressor with purely 2D blading exhibited extensive 3D separation regions at off-design conditions and was representative for the Rolls-Royce Trent series of civil aero engines in terms of aerodynamics. The focus of the application was on off-design conditions and the optimization was carried out at 15% surge margin. For the design study, only the hub end wall of the 3rd stator was individually optimized and the results again compared to the 2D and 3D airfoil. The best hub end wall shape was afterwards, the perturbation amplitudes being scaled on the corresponding chords, applied to the hub and casing surfaces of the remaining stator rows in the form of stereotypes. Concerning the rotor end walls, where large separations were not predicted, the end wall shape identified in the first part of the study was applied. Only the first stage rotor and the OGV were not modified since individual optimization was expected to be needed due to the highest Mach number and aerodynamic loading respectively. The application of contoured end walls improved the entire flow characteristics and increased surge margin by 3% compared to the 2D design. However, the level of the 3D case could not be achieved but, on the other hand, a slight increase in efficiency was observed. Harvey suggested to apply a leading edge re-cambering (LER) to the stators at the casing since they locally experienced a strong positive incidence due to the over tip leakage of the upstream rotor. This may lead to an increased calculated surge margin for a 2D design with profiled end walls and it could therefore be of particular benefit to include LER as an additional design feature in the optimization process.

The approach of 3-D CFD analysis of non-axisymmetric end walls at real engine conditions has been parallely followed up in the studies of Reising et al. [75, 74, 76, 77]. The redesign of the stator row of Configuration I of the Darmstadt Transonic Compressor with non-axisymmetric end walls is described in [76]. An approach of two independent optimizations of the hub and casing end walls considering two different operating points was pursued. This led to an 1.8% increase in efficiency at design conditions and remarkable improvements over the entire speed line. The follow-up work [77] deals with application of profiled end walls to the rotor hub end wall while investigating different design strategies. These reports also provide a comparison of how the PEW designs for stators and rotors differ - not just in terms of their shapes but also concerning the different effects on the flow fields. The latter two papers will serve as a principal base for the present design exercise which will be described in Chapter 4 and 5 of this thesis. As a preliminary study the hub and casing end walls of the stator had both been optimized at design conditions keeping the other end wall surface constant in each case in order to separate the effects for the root and tip section [74].

Both the studies of Harvey and Reising identified a mechanism that avoids or at least reduce separation and corner stall by enhancing the cross flow in the rear blade passage which forces the loss core to lift from the end wall surface and migrates adjacent to the airfoil suction surface towards mid-span. That way an accumulation of low-stagnation fluid can be prevented resulting in the observed huge loss reduction and the higher static pressure recovery. It could be shown that this mechanism was very similar to the effect of compound lean which is e.g. described by Denton and Xu [13]. This mechanism seems to be very powerful in suppressing corner stall and is quite different compared to the performance of profiled end walls in turbines where most of the benefits would be expected to result from reducing the static pressure gradient in the viscous flow region. However, secondary flow phenomena and corner separation in modern multi-stage compressors are normally kept under control by applying 3D design features. Hence, it is expected that retrospective integration of profiled end walls in already existing compressors may be the first opportunities for a successful application of this design feature, as has been successfully shown by Blackburn et al. [5] in the previously described redesign of an industrial turbine.

2.2.2 Application of Dihedral and Sweep

Sweep and dihedral, both terms having their roots in aircraft wing aerodynamics, are nowadays widely spread features applied to turbomachinery blades. An excellent overview on this topic is given by Denton and Xu [13] who laid down the principles of 3D flow effects in turbomachinery. Sweep and lean produce 3D effects and are therefore considered to be capable to positively influence secondary flow and to reduce the effect of stream surface twist due to large spanwise velocities within the blade row. Unfortunately, 3D effects can hardly be predicted by two-dimensional or quasi-three-dimensional blade-to-blade calculations as streamlines become twisted on passing through the blade row. Some approaches towards 3D design systems that have been used in industry for multistage compressor design in the recent years incorporate simple methods into the throughflow models to adapt 2D to 3D blade shapes. The problem is that these approaches depend on the appropriateness of the correlations introduced to allow for three-dimensionality. Thanks to the increasing availability of fully 3D flow field simulation methods, predictions of these complex effects are much faster available and more accurate so that they are now considered in the design process. Moreover, the use of CFD finally allows the coupling of blade rows and avoids the mismatching in the radial and axial directions [22].

Denton and Xu [13] provide a very general definition of sweep and lean. A blade is recognized to be swept if the leading edge is not perpendicular to the incoming flow or if the trailing edge is not perpendicular to the leaving flow respectively. In this context, positive sweep is defined by end wall sections moved parallel to the chordline towards the incoming flow. In contrast, a blade is considered as leaned if the intersection of the blade surfaces are inclined to the radial direction (i.e. end wall sections are moved perpendicular to the chord line), typically creating an obtuse angle between the end wall and the airfoil suction surface which is also known as positive dihedral [13]. This means that this feature is presented by differential displacement of airfoil sections up the span, perpendicular to the chord line. Although not all details of these complex effects are fully understood yet, a greater knowledge has been gained throughout the recent years by numerous studies at universities and in industry. To mention some of them, Gümmer et al. [30] presented with the application of sweep and dihedral to the fan stator of the BR710 the first approach in open literature that concerned both elements in an engine worthy airfoil design. Gallimore et al. [22] carried out a fundamental investigation into the effect of sweep and dihedral in a low speed single stage rig. With the help of the obtained findings, they developed a 3D design toolkit to design 3D bladings at the same aerodynamic duty as the corresponding 2D bladings. Finally, they introduced sweep and lean to the design of the core compressor for the Rolls-Royce Trent engine series [23] with the target to improve efficiency without reducing the stability margin. Similar to the application of non-axisymmetric end walls, the retrospective application of 3D effects to already existing bladings is also an issue of high interest.

Effects of Sweep

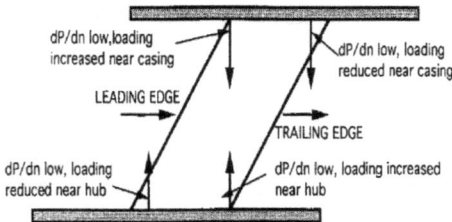

Figure 2.7: Effect of sweep on blade loading taken from Denton and Xu [13]

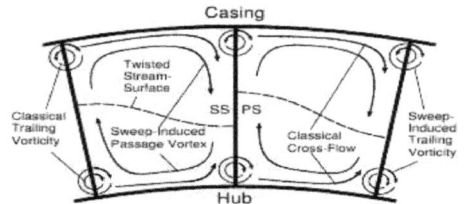

Figure 2.8: Classical and sweep induced secondary flow structures from Gümmer et al. [30]

When sweep is applied, the axial velocity component will increase on the suction side and decrease on the pressure side respectively as the flow passes through a blade row. Thus, streamlines close to the suction side will have a lower pitch angle than those close to the pressure side as the spanwise velocity components remain constant. As a result, the stream surfaces become twisted with the connected impact on blade loading to be observed near the end wall. As described in Gümmer [30], this non-axisymmetric stream surface twist generates additional secondary flow features in the passage. A large vortex will be created across the entire blade channel as shown in Figure 2.8. This vortex is counteractive to the classical cross flow at the hub but amplifying the cross passage flow near the casing. Close to the end walls, the streamline off-set disappears and additionally vortex structures are shed from the trailing edge leading to a much more complex secondary flow pattern in a swept blade than in a conventional one. A different approach to the effect of sweep is made by considering its influence on blade loading.

Figure 2.7 shows an example with positive sweep on the lower end wall and negative sweep on the upper end wall (in general, sweep and dihedral angle are defined at the pinch point of the stacking axis and the corresponding stream surface). The effect can be understood by considering that the pressure gradient perpendicular to the end wall is much smaller compared to the blade-to-blade pressure gradient. Since there is no blade above the end wall sections at the leading edge to support the pressure rise that would be generated if the flow was purely two-dimensional, the blade loading is reduced because the blade force has to fall to zero shortly above the end wall sections. This leads to a redistribution of the loading. The hub region at the leading edge is unloaded and the entire hub section aft-loaded. The tip region experience the opposite and is front-loaded. This is of special importance for compressors where the loading at the leading edge tends to be much higher than in turbines which may help to avoid early blade stall or corner separations. Gallimore et al. [22] observed in their investigations on the Deverson compressor that the application of negative sweep at the trailing edge even caused a separation to occur at this position due to overloading. Thus, it would be worthwhile within design process to consider the possibility to apply positive sweep on both the upper and the lower end wall.

If we recall what has been mentioned about the correlation between aft-loaded profiles and the intensity of secondary flow in the previous chapters the application of forward sweep should reduce secondary flow and additionally make the leading edge more tolerant to changes in incidence and boundary layer skewness. Apart from that, sweep might also have an effect on shock losses of transonic compressors and fans by inclining the shock pattern. In fact, the first intention of the application of sweep was to reduce shock losses analogous to swept wings for aircrafts. As the shock loss depends on the Mach number component perpendicular to the shock front [12], these losses can be decreased by any sweep of the shock and theory would suggest that this could be achieved either by positive or negative sweep. This effect of sweep was investigated in detail in a study of Wadia et al. [94] who analyzed the influence of forward and backward sweep on the compressor stall margin in a highly-loaded military fan. According to their study backward sweep may suffer a loss in stall margin as the shock might move further to the leading edge where as, conversely, the swept forward blade showed an increased surge margin with a shock located closer to the trailing edge. However, they found the efficiency of conventional, back and forward swept fans to be very similar. The flow through a fan that is designed for the same duty must experience the same pressure rise for any of the three differently swept fans. Near the tip, the shock produces most of the pressure rise which means that if the strength of the shock is reduced by sweeping the shock relative to the flow direction, then a greater part of the pressure rise must result from the subsonic diffusion for compensation. This will increase the boundary layer losses and eventually even lead to separation. Hence, an overall gain in efficiency is not obvious since the reduced shock loss may be outweighted.

These findings were followed up in a later study by Denton and Xu [14] who investigated the interaction of 3D features with the shock pattern in a series of different designs. They, in turn, report on a very little effect of sweep and lean on the shock pattern near the tip where the shock has to be perpendicular to the casing again. However, they also observed the positive effect on the stall margin when forward sweep was applied but found out that the effect of sweep itself was hardly responsible for the improvements. For some designs the bow and passage shock could be observed as almost completely merged to form one single stronger shock located immediately behind the leading edge. The lack of any pre-compression from a separated bow shock resulted in lower efficiency levels compared to the conventional design. The authors conclude that the reduction in overall shock loss is much more effective by realizing the compression by two shocks rather than by one single one. Therefore, the bow and passage shocks must remain distinct and should not merge into one single structure. Moreover, for their fan the application of forward sweep and lean reduced the choking flow while backward sweep and lean did the opposite. This effect was observed to be proportional to axial displacement of the tip sections not depending on whether produced by sweep or lean which resulted in a change in annulus area and throat area. Additionally, the application of sweep and lean had an effect on the interaction between bow and passage shock (not to be mistaken for the shock inclination!).

Effects of Dihedral

The application of lean is characterized by a non-radial blade stacking as as illustrated in Figure 2.9. Considering a hypothetical two-dimensional flow there would be iso-pressure surfaces aligned with the leaned blade. This would lead to a pressure gradient perpendicular to the end walls involving a radial acceleration of the flow and therefore velocity components towards mid-span. This effect is a very powerful tool to keep low-momentum fluid away from the end wall, especially if compound lean is applied (bowed blade at hub and shroud). The low stagnation fluid in the blade and end wall boundary layer will migrate towards mid-span more strongly compared to conventional airfoils. It is assumed that this effect might avoid an accumulation of low energetic fluid and therewidth separation in the end wall region.

Nonetheless, one of the most common purpose of blade lean is to change the spanwise variation of the blade reaction leading to radial inward forces. This is possible because dihedral adds an additional force to the radial equilibrium of the flow between hub and casing. Denton and Xu [13] derived an equation of this force depending on two terms. One term incorporates change in streamline curvature whereas the second one the change in radial pressure gradient. Concerning streamline curvature, dihedral reduces the induced velocities on the profile suction side and increases them on the pressure side. All over this leads to a reduction in peak Mach number and suction side diffusion as observed in the comparisons of Gümmer et al. [30] between conventional and advanced fan stator profiles. In contrast to sweep, dihedral has no influence on the classical secondary flow structure [30]. For high aspect ratios the streamline curvature term dominates and the change in the radial pressure gradient tends to be comparatively small. However, the streamline curvature has an effect on the pressure upstream and downstream of the blade row since streamline curvature cannot end immediately at the leading and trailing edges. Similar to a bump applied in the context of non-axisymmetric end wall contouring this stream line curvature leads to an increased static pressure upstream and downstream of the blade channel.

There are different explanations that have been put forward to argue how blade lean affects the end wall flow and might lead to improvements.According to Denton and Xu [13], the effect of different blade stacking can be compared with moving the blade within an almost frozen

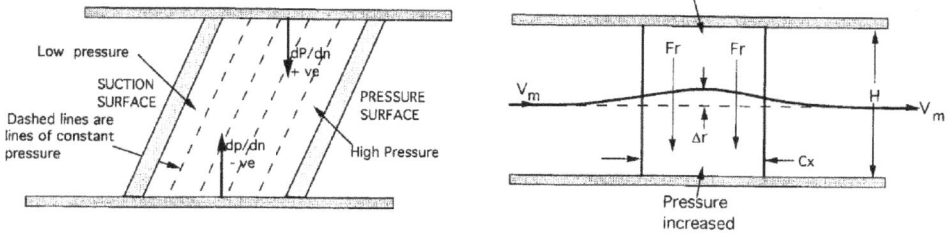

Figure 2.9: Pressure iso-surfaces in a hypothetical leaned blade and the effect of lean on streamline curvature taken from Denton and Xu [13]

pressure field. From this point of view, straight lean inceases tip loading and reduces hub loading whereas bowed or compound lean relieves both hub and tip sections at the expense of a higher load at mid-span which is often considered as the most efficient part of the blade. Negative dihedral may even lead to corner separation due to the increased blade loading involving additional blockage and thus higher losses which was demonstrated by Gallimore et al. [22].

Nonetheless, this argumentation is not fully unambiguous. Some researchers concluded that the application of bow rather results in a radial redistribution of loss than a real reduction. Harrison et al. [36] investigated 3 different highly loaded, low aspect-ratio turbine blades (unleaned, straight lean and compound lean) in a cascade. He found a marked effect of lean on blade loading upon loss distribution and the state of the boundary layer on the suction side of the blade and the end walls. However, he did not observe a loss reduction in the blade row, where it was applied. He even reported, opposed to the loss flux, on higher local loss coefficients at mid-span while the coefficients were found to be reduced near the end walls. The higher loss at mid-span seemed to appear as a consequence of the higher blade loading and because the spanwise pressure gradient along the blade suction surface led to a greater accumulation of low-momentum fluid at mid-span. The straight leaned blade showed even a slightly higher mass-averaged loss coefficient. A redesign of the mid-span blade sections to tolerate the increased loading should result in a total pressure loss decrease for the entire row. In fact, considering the mid-span sections to remain unchanged, the overall improvement which could be observed for the bowed blade originated from the reduced mixing losses downstream of the trailing edge.

The compound leaned blade row gives less mixing loss than a conventional blade because more of the loss is already concentrated at mid-span and therefore the mixing between the flow in the boundary layer and the flow at mid-span is less intense. Moreover, the application of lean proved to compensate the over- and underturning of the passage vortex providing less incidence variation and improved inflow conditions for the following downstream stages. A second explanation that is often cited is the aforementioned redistribution of the low energy fluid connected to less concentrated secondary loss cores. Amongst others, Gallimore et al. [23] report on eliminating areas of reverse flow in a stator blade row at the design point and significantly reducing them near stall by applying positive lean. A third explanation is based on the entropy production per unit surface area that is proportional to the local relative velocity.

A smaller velocity results in less entropy production [12]. When compound lean is applied the area of surface exposed to a decreased velocity (end walls, root and tip sections of the blade) is higher than the area of surface exposed to a greater velocity (mid-span sections), Therefore, this should lead to a diminution of the overall entropy production.

Potential and Limits of Sweep and Dihedral

The previous chapters have shown the potential of sweep and dihedral to control and improve the three-dimensional end wall flow. While sweep mainly serves to influence the chordwise loading distribution, dihedral primarily can be used to modify spanwise loading distribution while maintaining the style of the profile's static pressure distribution at the same time. Nonetheless, sweep also impacts the radial pressure distribution to some degree as an off-loading of the end walls can lead to an increased loading over the rest of the span, especially in regions where the dihedral or sweep blends back into the conventional stacking line. This may attract the migration of boundary layer fluid and care must therefore be taken concerning the stacking line shapes to avoid overloading of any blend point by a combination of the increase in loading that is necessary to compensate the reduced end wall loading. Gallimore et al. [23] report on this issue when carrying out a series of design exercises. One modification consisted of the redesign of a conventional rotor-stator combination. For both blades, axial and tangential shifts were allowed. The final design of the stator blade showed both sweep and dihedral at both ends and a significantly reduced chord length at mid-span. This led to the interesting result that the most severe regions of reverse flow on the stator suction side were predicted to occur away from the end walls. Concerning the overall characteristics, this led to a slight reduction in the predicted stall range compared to the conventional design which was attributed to the mid-span chord reduction in an attempt to maximize efficiency. The experimental stall range later proved to be similar to the datum one as the stalling process for the investigated compressor was dominated by the end wall regions which could not be captured by the CFD. However, sharp transitions between the end wall portion with positive sweep or positive dihedral and the conventionally stacked blade region should be avoided in this regard, since they behave similar to modifications with negative sweep or dihedral generally involving increased aerodynamic loading and loss [30].

A limitation of sweep, although not of aerodynamic nature at first sight, may also be due to the mechanical integrity in the entire engine. Sweep can lead to problems in matching for multistage configurations if the axial gaps between adjacent blade rows and the length of the whole compressor should or must be maintained. This applies in particular when blade or vane rows are retrofitted within an already existing global design as shown in the high-speed testings on a six-stage high-pressure compressor by Gallimore et al.[23] who had to restrict the axial movements to maintain acceptable gaps between the different blade rows. Their findings were used to design the intermediate and high pressure compressors for the Rolls-Royce Trent 500 engine including 3D features from the very beginning which suggests that sweep and lean will be key elements of all future compressor design developments.

2.2.3 Leading Edge Modification and End Wall Fences

In this subchapter a brief overview on the fields of leading-edge modification and end wall fences will be given. Both are alternative passive methods which have a favorable influence on the end wall flow while they are based on different mechanisms than non-axisymmetric end walls. Leading-edge modifications consists either of fillets or bulbs and primarily focus on affecting the formation of the horseshoe vortex in the leading-edge region to positively alter the entire vortex system in the blade channel. In contrast, the purpose of end wall fences is to guide the trajectory of the cross flow as well as the passage vortex through the blade channel on an intended way to minimize the amount of low-momentum flow that impinges on the blade suction side.

Quite a lot of the studies on leading-edge modification that can be found in literature are from Sauer and Müller [82, 65, 66]. Based on the idea that the generation of secondary flow is greatly affected by the horseshoe vortex formation, they applied a bulb on a turbine airfoil that had been developed for a highly loaded low pressure turbine [82]. A bulb has a rectangular junction with the end wall which rises in spanwise direction before it merges tangentially into the profile by some trigonometrical function. In contrast, fillets have a tangential intersection (or at least with a very small angle) both with the end wall and the blade surface. The expectation of this study was that the application of the bulb would intensify the suction side leg of the horseshoe vortex. The suction side leg would keep the passage vortex away from the suction side profile boundary layer by its opposite rotational direction and weaken its effect. In order to isolate secondary loss the profile loss was subtracted. The authors reported on a 50% reduction in secondary loss compared to the reference case whereas the numerics were able to reproduce the experimental findings. The loss decrease could especially be found in the areas close to the end wall and at around 20% span. This area coincidenced with the extension of the passage vortex causing over- and under-turning respectively. Some first guidelines how the design of a leading-edge should look like could be derived. The optimal geometry turned out to be a non-symmetric bulb consisting of a pronounced suction side and a less extended pressure side although variations of the boundary layer and incidence led to different dimensions of the bulb. This work was continued by Müller et al. [65] whose objectives were basically the same unless his experimental and numerical work was focused on a compressor cascade. He applied similar bulbs in a cascade with a low turning airfoil of $18°$. Based on the same interaction, an isolated secondary loss reduction of approximately 20% could be obtained at design condition. The smaller decrease compared to the turbine cascade is not surprising due to the much lower turning and therewidth smaller resulting secondary flow. Because compressor blades suffer a high sensitivity to incidence the whole working range of the compressor was analyzed, especially the area near stall. At very high incidence angles the bulb failed to work properly due to the strong separation which occurred in the blade corner. The authors concluded that a further improvement without other means would not be possible at high loadings near stall.

The investigations of leading-edge modification in compressors was extended on fillets by the same group of researchers. In [66], a comparison of 5 different end wall modifications in a compressor cascade was presented which consisted of a bulb, a medium and a large fillet. The large fillet had twice the radius of the medium one and both fillets were modified by an axial blunt cut-off at the leading edge to represent the case of very small gaps between to rows in a multi-stage compressor. In addition to what had already been known from the previous studies, they observed the rise of a new fillet vortex rotating in the same direction as the passage vortex for the fillet and blunt-fillet configurations. At design conditions, the bulb was observed to reduce loss until 15% span and moved the maximum loss area towards the end wall. The large fillet led to an overall slight loss reduction. However, the axial blunt cut-off seems to produce additional losses as they increased dramatically for the blunt leading edge of the large fillet. The medium fillet resulted in a mitigated loss close to the end wall coevally yielding a higher level at 5-10% span which compensated each other. The blunt version of the medium fillet showed more or less the same behavior. Further investigations indicated that the blunt leading edge of the large fillet radius caused a stable separation close to the end wall leading to the dramatic rise of isolated secondary loss by 24%. In contrast, a blunt leading edge with the smaller fillets did not result in a separation but produced a horseshoe-like vortex. Theses results show the sensitivity of such a system to small geometrical changes. The benefits of the intended stronger suction side leg of the horseshoe vortex can be negated by additional loss features which occur due to the new features. The highest reduction of secondary loss with 28% was observed for the application of a bulb at an incidence of 3°. At design conditions the authors reported on an improvement of net end wall losses from 7 up to 17% depending on the applied feature.

Hoeger et al. [51] investigated nominal and high incidences for different fillet radii to explore the possibility to extent compressor loading limits and the operating range since the surge margin of a compressor might also be influenced by the stator hub and tip region. In this context, the leading edge shape an the presence of a fillet may be particular importance. Apart from basic experiments a high number of numerical simulations with differently skewed and distorted inlet boundary layers were carried out. Generally, the fillet configurations allowed higher incidences before an onset of corner stall could be observed. At nominal swirl, the blade loading of the datum cascade showed increasing Mach number levels and less diffusion towards the trailing edge on sections near the end wall region which indicated a corner separation. This corner separation could be avoided by the fillet. However, the fillet had no influence on the mid-span flow in absence of boundary layer twist. For the highly skewed boundary layer this changed completely as the reference cascade could no longer withstand the high incidences at the end wall. This led to a complete separation along the profile which the fillet was able to repair.

More recently, Becz et al.[2] performed wind tunnel testing on on two promising designs of the bulb geometry which had been introduced by Sauer and on a leading edge fillet. The features were applied to a low aspect ratio high-turning airfoil configuration in a large scale low speed turbine cascade. Based on the evaluation of the total pressure loss coefficients the large bulb was unable to reduce total loss when compared to the reference airfoil. On the other hand, the small bulb and the fillet radii worked very well leading to a significant reduction of 8% for both design. However, the authors cautioned that these results did not represent a complete picture of the end wall loss decrease because the measured values were area averaged. Therefore, an exact representation of the entropy production could not be captured. To overcome this difficulty, Becz et al. [3] carried out a second set of experiments to obtain mass-averaged values of the total pressure loss. Only the two designs that had demonstrated an area-averaged loss reduction in their first study and the reference airfoil were considered, i.e. the large bulb was not included. The same test facility was used upgraded by the possibility to measure mass flux. The received results were quite surprising. In contrast to the area-averaged measurements the small bulb did not show any loss reduction for the mass-averaged values which emphasizes the challenge to select the proper averaging method. Yet, the application of the bulb involved a slightly higher turning which is an indication for an increased airfoil loading but without the additional loss penalty that is normally observed. The fillet, in contrast, reduced total pressure loss in agreement to the first study by approximately 7% while the blade loading was slightly decreased. The fillet radius changed the size of the passage vortex pattern remarkably whereas no change of the pattern was observed for the bulb configuration. In agreement to theory, the suction side leg of the horseshoe vortex was strengthened by applying the bulb which produced higher loss at around 30% span. In other words, the reduced interaction between horseshoe and passage vortex did not seem to be significant enough to overcompensate the increased loss for the tested geometry. Both modifications yielded a reduction of the extent of the loss area in the near end wall location resulting in higher end wall loading and increased overturning. This overturning was more pronounced for the bulb configuration resulting in the mentioned blade load increase but slightly negated by an underturning in the mid-span sections for the fillet. Through this second set of experiments it was shown that the area of greatest loss reduction differed between the bulb and the leading edge fillet which suggested that the mechanisms through which each influenced the loss production may be different. This fact might offer the potential to combine the advantages of both features in the future.

Working with a compressor cascade with CDA blades, Zhong et al. [96] studied the effect of end wall fences on the end wall boundary layer. The application of end wall fences affects the near end wall flow in two ways. The cross flow from pressure surface to suction surface is obstructed as the streamlines impinges further downstream onto the surface of the fence. The exact impact depends on the pitchwise position of the fence. This leads to the formation and development of a fence vortex which lifts along the end wall fence. The fence vortex rotates in the opposite direction of the passage vortex and is therefore counteracting the secondary flow structures leading to additional blockage. In general, stronger cross flow induces a stronger fence vortex. The second effect is that the end wall fence can be seen as a kind of an additional (splitter) blade involving the same complex vortex system like an airfoil with an additional horseshoe vortex. Consequently, the end fall fence can be considered to split the blade channel into two smaller passages each having its own secondary flow system. The additional vortices will lead to an extra loss in the entire passage. Hence, the goal for such configurations is to find an optimum between the extra loss and the expected positive influence of the new counterrotating vortex structures on the size and strength of the passage vortex. which is quite similar to the application of bulbs. The optimal fence geometry depends on many parameters such as length, pitchwise position, thickness and height, the height being expressed in relation to the incoming boundary layer. Zhong et al. [96] run a numerical study on theses parameters. They found their best sample with a height of one third of the incoming boundary layer, 75% of axial chord length and at 30 % pitch from the pressure side. They observed that increasing the height and length further led to an even higher blockage effect which is desirable. However, this was compensated by the additional loss produced by the fence itself. Moreover, the blocking effect depended on the pitch location of the fence. This means that an end wall fence works properly if the two vortex systems are smaller than the original bigger one in the datum passage. With that technique the authors were able to reduce the cascade loss by 7-9%.

3 Governing Equations and Design Principles

This chapter gives an overview of the relevant governing equations together with a brief introduction into the principles of computational fluid mechanics (CFD). In the last part of this chapter the used optimization scheme will be provided.

3.1 Basic Equations of Fluid Mechanics

In this section the conservation equations, which are relevant for the computation of fluid flows, are briefly introduced. A more detailed description can be found in [89], [47] and [48].

3.1.1 Conservation of Mass

The mass m of a body or a volume of fluid can be expressed as the integral of its density ϱ over the volume V:

$$m = \iiint_{V(t)} \varrho(\vec{x}, t)\mathrm{d}V \tag{3.1}$$

where density ϱ is a function of space $\vec{x}$ and time t. Because mass is neither produced nor annihilated, i.e.

$$\frac{\mathrm{D}m}{\mathrm{D}t} = \frac{\partial m}{\partial t} + u_i \frac{\partial m}{\partial x_i} = 0 \tag{3.2}$$

equation (3.1) can be written as

$$\frac{\mathrm{D}}{\mathrm{D}t} \iiint_{V(t)} \varrho \mathrm{d}V = \iiint_{V} \left[\frac{\partial \varrho}{\partial t} + \frac{\partial}{\partial x_i}(\varrho u_i)\mathrm{d}V \right] = 0. \tag{3.3}$$

The operator $\frac{\mathrm{D}}{\mathrm{D}t}$ is known as the material derivative. Since the integrand of this expression must be zero, the following expression can be concluded

$$\frac{\partial \varrho}{\partial t} + \frac{\partial}{\partial x_i}(\varrho u_i) = 0 \tag{3.4}$$

which is known as the *continuity equation*.

3.1.2 Momentum Conservation Equations

The change of momentum must be equal to the integral over all forces acting on a single fluid element. These forces are divided into body forces $\vec{k}$ which act on the volume V of a body and surface tensions $\vec{t}$ which act on the surface S of a body. This leads to

$$\frac{D}{Dt} \iiint_{V(t)} \varrho \vec{u} \, dV = \iiint_V \varrho \vec{k} \, dV + \iint_S \vec{t} \, dS \tag{3.5}$$

with $\vec{k} = \lim_{\Delta m \to 0} \frac{\Delta \vec{F}}{\Delta m}$ and $\vec{t} = \lim_{\Delta S \to 0} \frac{\Delta \vec{F}}{\Delta S}$. The vector $\vec{t}$ can also be expressed in terms of a stress tensor T as $\vec{t} = \vec{n} T$ with the stress tensor being

$$T = \begin{pmatrix} \tau_{ii} & \tau_{ij} & \tau_{ik} \\ \tau_{ji} & \tau_{jj} & \tau_{jk} \\ \tau_{ki} & \tau_{kj} & \tau_{kk} \end{pmatrix} \tag{3.6}$$

Substituting this into equation (3.5) and changing the surface integral into a volume integral by Gauss' integral theorem the the following expression is determined:

$$\iiint_V \left(\varrho \frac{Du_i}{Dt} - \varrho k_i - \frac{\partial \tau_{ij}}{\partial x_j} \right) dV = 0 \tag{3.7}$$

From here the differential form of the conservation equations of momentum is derived:

$$\varrho \frac{Du_i}{Dt} = \varrho k_i + \frac{\partial \tau_{ij}}{\partial x_j} \tag{3.8}$$

3.1.3 Conservation of Energy

The change of energy with time is caused by the work of external forces and by energy introduced into the system as stated in the first law of thermodynamics. The energy of a system consists of the internal and kinetic energy. External forces are volume and surface forces $\vec{k}$ and $\vec{t}$ as explained in the above section 3.1.2. Supplied energy in terms of heat is denoted as q. Thus, the following equation is obtained:

$$\frac{D}{Dt} \iiint_{V(t)} \left[\frac{u_i u_i}{2} + e \right] \varrho \, dV = \iiint_V u_i k_i \varrho \, dV + \iint_S u_i t_i \, dS - \iint_S q_i n_i \, dS. \tag{3.9}$$

Changing the surface integrals into volume integrals and defining enthalpy as $h = e + \frac{p}{\varrho}$ the commonly used form of the energy equation is derived:

$$\varrho \frac{D}{Dt} \left[\frac{u_i u_i}{2} + h \right] = \frac{\partial p}{\partial t} + \varrho k_i u_i + \frac{\partial}{\partial x_j} (P_{ij} u_i) - \frac{\partial q_i}{\partial x_i} \tag{3.10}$$

where P is a tensor of friction forces.

3.1.4 Navier-Stokes and Euler Equations

For Newtonian fluids the stress tensor τ_{ij} can be related to material properties η and λ^*, which are functions of the thermodynamic state, i.e. $f(p, T)$.

$$\tau_{ij} = -p + \lambda^* \frac{\partial u_k}{\partial x_k} + \eta \left[\frac{\partial u_i}{\partial x_j} + \frac{\partial u_j}{\partial x_i} \right] \tag{3.11}$$

Substituting this relation into the conservation of momentum (3.8) the *Navier-Stokes equations* are found as follows:

$$\varrho \frac{Du_i}{Dt} = \varrho k_i + \frac{\partial}{\partial x_i} \left\{ -p + \lambda^* \frac{\partial u_k}{\partial x_k} \right\} + \frac{\partial}{\partial x_j} \left\{ \eta \left[\frac{\partial u_i}{\partial x_j} + \frac{\partial u_j}{\partial x_i} \right] \right\} \tag{3.12}$$

For inviscid, non-conducting flows, i.e. $\eta, \lambda^* = 0$, the Euler equation is obtained:

$$\varrho \frac{Du_i}{Dt} = \varrho k_i - \frac{\partial p}{\partial x_i} \tag{3.13}$$

The continuity equation (3.4), the conservation of momentum (3.5) and energy (3.10) can be arranged in one vector equation which finally leads to the following nonlinear system.

$$\frac{\partial}{\partial t} \begin{pmatrix} \varrho \\ \varrho u \\ \varrho v \\ \varrho w \\ \varrho E \end{pmatrix} + \frac{\partial}{\partial x} \begin{pmatrix} \varrho u \\ \varrho u^2 + p \\ \varrho uv \\ \varrho uw \\ \varrho Eu + pu \end{pmatrix} + \frac{\partial}{\partial y} \begin{pmatrix} \varrho v \\ \varrho uv \\ \varrho v^2 + p \\ \varrho wv \\ \varrho Ev + pv \end{pmatrix} + \frac{\partial}{\partial z} \begin{pmatrix} \varrho w \\ \varrho uw \\ \varrho vw \\ \varrho w^2 + p \\ \varrho Ew + pw \end{pmatrix} = 0 \tag{3.14}$$

3.1.5 Rotating Frame of Reference

In turbomachinery flows, we have to deal with a rotating frame of reference and it is therefore necessary to describe the conservation laws relatively to a rotating frame of reference [47]. Defining $\vec{w}$ as the relative velocity field and $\vec{u} = \vec{\omega} \times \vec{r}$ as the entrainment velocity, the absolute velocity consists of

$$\vec{v} = \vec{w} + \vec{u} = \vec{w} + \vec{\omega} \times \vec{r}. \tag{3.15}$$

The entrainment velocity does not contribute to the mass balance, hence, the continuity equation in the relative system can be written as

$$\frac{\partial \varrho}{\partial t} + \frac{\partial}{\partial x_i} (\varrho w_i) = 0. \tag{3.16}$$

With regard to the momentum conservation law, the observer in the rotating frame of reference will see two additional forces compared to the absolute frame of reference. These are the Coriolis force per unit mass

$$\vec{f_c} = -2 (\vec{\omega} \times \vec{w}) \tag{3.17}$$

and the centrifugal force per unit mass

$$\vec{f}_c = -\vec{\omega} \times (\vec{\omega} \times \vec{r}) = \omega^2 \vec{r} \tag{3.18}$$

where $\vec{R}$ is perpendicular to the axis of rotation. These additional force terms have to be added in the right hand side of the momentum conservation equation if this equation is directly written in the rotating frame of reference. The conservation law for momentum in the relative system then becomes

$$\frac{\mathrm{D}}{\mathrm{D}t} \iiint_{V(t)} \varrho \vec{u} \mathrm{d}V = \iiint_V \varrho \vec{k} \mathrm{d}V + \iiint_V \varrho \vec{f}_C \mathrm{d}V + \iiint_V \varrho \vec{f}_c \mathrm{d}V + \iint_S \vec{t} \mathrm{d}S \tag{3.19}$$

in which the shear stress tensor derived in equation 3.6 is to be expressed as function of the relative velocities. In order to obtain the energy conservation law in the relative frame of reference, the work of the centrifugal needs to be added as the Coriolis forces do not contribute to the energy balance of the flow. For a comprehensive derivation of the energy equation in the relative system, it is referred to [47] and [48].

3.2 Turbulence Models

Turbulent flows are of unsteady, fluctuating nature which are characterized by non-deterministic changes of all flow variables. Rotation is unequal to zero and energy is transported from large to small scales, until it is dissipated due to viscous shear forces. Turbulence is generated above a critical Reynolds number that may range range from 400 to 3000 depending on the specific case. In most industrial applications the Reynolds number lies above that range. Turbulent flows contain an increased level of friction which will lead to higher losses. On the other side, turbulence can also have a beneficial influence on the flow in terms of a better mixing behavior or re-energizing the boundary layer to avoid separation. In order to predict adequately the turbulence effects on the flow field all variables are decomposed into a constant and a fluctuating value

$$\phi(x_i, t) = \overline{\phi(x_i)} + \phi'(x_i, t) \tag{3.20}$$

with the constant part being calculated by

$$\overline{\phi}(x_i) = \lim_{T \to \infty} \frac{1}{T} \int_0^T \phi(x_i, t) \mathrm{d}t. \tag{3.21}$$

The flow variables of the governing equations are expressed according to this approach and substituted into the Navier-Stokes Equations (3.12). Averaging of their values according to Favre [19, 20] finally leads to the Reynolds-Averaged Navier-Stokes Equations:

$$\frac{\mathrm{D}\overline{u_i}}{\mathrm{D}t} = \nu \nabla^2 \overline{u_i} - \frac{\partial \overline{u_i' u_j'}}{\partial x_i} - \frac{1}{\varrho} \frac{\partial \overline{p}}{\partial x_i} \tag{3.22}$$

The only difference between these equations and the original Navier-Stokes equations is the Reynolds stress tensor $\overline{u_i' u_j'}$. Conservation equations for this term contain further unknowns of higher order. This generates an infinite hierarchy of differential equations what is called *closure problem*. Different models are developed to resolve these equations. Unknown terms are modeled or approximated using given variables. Depending on the complexity and the number of additional differential equations to be solved, these models can be divided into three main kinds:

- algebraic models (e.g. Baldwin-Lomax)

- one-equation models (e.g. Spalart-Allmaras)

- two-equation models (k-ϵ, k-ω)

For this work, the one-equation model by Spalart-Allmaras [88] has been applied in all presented test cases.Therefore this section is limited to a brief introduction of this model.

3.2.1 The Spalart-Allmaras Model

The one-equation turbulence model of Spalart-Allmaras can be seen as a bridge between the algebraic and the two-equation models. This model has become very popular over the last years due to its robustness and its ability to treat complex flows. The main advantage of the Spalart-Allmaras model is the continuous turbulent eddy viscosity field compared to the algebraic models and the lower additional CPU cost compared to models of higher order. It was originally developed for external aircraft wing aerodynamics. Since turbomachinery blades can be considered as small highly-loaded 'wings', this model is nowadays widely spread in the turbomachinery sector.

The principle of this model is based on the resolution of an additional transport equation for the eddy viscosity. The equation contains an advective, a diffusive and a source term. For this model, a turbulent working variable (Spalart-variable) $\tilde{v}$ obeys the transport equation

$$\frac{\partial \tilde{v}}{\partial t} + u_j \frac{\partial \tilde{v}}{\partial x_j} = C_{b1} \left[1 - f_{t2} \right] \tilde{S} \tilde{v} + \frac{1}{\sigma} \left\{ \nabla \cdot \left[(v + \tilde{v}) \nabla \tilde{v} \right] + C_{b2} \left| \nabla v \right|^2 \right\} \tag{3.23}$$

$$- \left[C_{w1} f_w - \frac{C_{b1}}{\kappa^2} f_{t2} \right] \left(\frac{\tilde{v}}{d} \right)^2 + f_{t1} \Delta U^2.$$

and is connected to the turbulent viscosity by

$$v_t = f_{v1} \tilde{v}. \tag{3.24}$$

f_{v1} describes a damping function defined by

$$f_{v1} = \frac{\chi^3}{\chi^3 + C_{v1}^3} \tag{3.25}$$

with χ being the ratio between the Spalart-variable $\tilde{v}$ and the molecular viscosity v. The production portion of the Source term is constructed with the following functions:

$$\tilde{S} = S f_{v3} + \frac{\tilde{v}}{\kappa^2 d^2} f_{v2} \tag{3.26}$$

with

$$f_{v2} = \cfrac{1}{\left(1 + \frac{\chi}{c_{v2}}\right)^3} \quad \text{and} \quad f_{v3} = \frac{(1 + \chi f_{v1})(1 - f_{v2})}{\chi};$$ (3.27)

where d is the distance to the closest wall and S the magnitude of vorticity. In the destruction term, the function f_w is

$$f_w = g \left(\frac{1 + C_{w3}^6}{g + C_{w3}^6}\right)^{\frac{1}{6}}$$ (3.28)

with

$$g = r + C_{w2}(r^6 - r) \quad \text{and} \quad r = \frac{\tilde{v}}{\tilde{S}\kappa^2 d^2}.$$ (3.29)

The constants arising in the model have been empirically determined with following values:

$$C_{w1} = \frac{C_{b1}}{\kappa^2} + \frac{1 + C_{b2}}{\sigma}, \quad C_{w2} = 0.3, \quad C_{w3} = 2, \quad C_{v1} = 7.1, \quad C_{v2} = 5,$$

$$C_{b1} = 0.1355, \quad \sigma = \frac{2}{3}, \quad C_{b2} = 0.622, \quad \kappa = 0.41$$

3.3 Numerical Methods

3.3.1 Computational Meshes

In many computational applications structured multi-block meshes are used to discretize the domain. These meshes consist of quadrilateral cells in 2D or hexahedral cells in 3D which are regularly connected. In most cases structured meshes can be applied resulting in low memory requirements and good accuracy in boundary layer resolution. For more complex geometries such as cooling holes in turbine applications unstructured meshes might be necessary. Any kind of element shape which the solver is able to handle can be used. The domain can also be divided into areas of structured and unstructured cells what is called a hybrid mesh.

Meshing Guidelines

This section gives an overview of meshing criteria which are to be satisfied to ensure a high mesh quality:

- *Orthogonality:* Skewed cells complicate numerical integration, the lower limit may depend on the solver. For the used software package of NUMECA a minimum angle of 30°should be obtained.

- *Near Wall Resolution:* Use at least 10 cells normal to the wall within a boundary layer. The non-dimensional wall distance y^+ is defined as follows:

$$y^+ = \frac{u_* y}{\nu} \qquad (3.30)$$

with the friction velocity $u_* = \sqrt{\frac{\tau_w}{\varrho}}$. The applied low Reynolds number model of Spalart-Allmaras does not incorporate wall functions which means that the boundary layer needs to be completely resolved. Therefore higher values than $y^+ = 10$ should be avoided. For models with wall functions $y^+ \approx 20 \div 50$ is sufficient.

- *Expansion factor:* A factor of 3 at the maximum should be intended to avoid abrupt changes in cell size.

- *Aspect ratio:* An approximate range for the ratio of cell length and width should be lower than 1000.

These criteria has been taken as a reference to judge the mesh quality of all cases throughout this work.

3.3.2 Explicit and Implicit Solvers

The governing equations of the fluid flow are discretized in space and time. Differential equations of the general form $\frac{\partial \varphi}{\partial t} = F(\varphi)$ need to be integrated over every time step Δt. Here, φ is an arbitrary scalar and the right hand side includes all spatial derivatives. A first order *explicit* Euler discretization reads as follows:

$$\frac{\varphi^{n+1} - \varphi^n}{\Delta t_n} = F(\varphi^n) \qquad (3.31)$$

where φ^n and φ^{n+1} are values at time step t and $t + \Delta t$. Using an explicit method, the value of the future time step is calculated from existing values according to the following equation:

$$\varphi^{n+1} = \varphi^n + \Delta t_n F(\varphi^n) \qquad (3.32)$$

The time step size Δt depends on the finest grid cell of width Δx for stability reasons. This is expressed in the Courant-Friedrichs-Lewy condition (CFL condition) $\frac{u \Delta t}{\Delta x} < C$. The constant C depends on the kind of equations to be solved. A global time step is used which is the minimum of all local time steps.

In contrast, the *implicit* Euler method evaluates φ^{n+1} using $F(\varphi)$ at the future time level as follows:

$$\varphi^{n+1} = \varphi^n + \Delta t F(\varphi^{n+1}) \qquad (3.33)$$

This equation is solved iteratively with an inner marching variable which is driven to zero for every physical time step. That is why the implicit method is also referred to as dual time stepping. It is unconditionally stable for any time step size. In general, implicit solvers are more costly because the demand of memory and computational time per time step is higher. This drawback is usually compensated because the time step size can be much larger without causing instabilities. Furthermore, the equations of implicit methods resemble those of steady state problems. Thus, the same solver can be used for both steady state and implicit transient computations.

3.3.3 Time Discretization

In turbomachinery applications, there are two common ways to approach an unsteady simulation in a time-efficient manner. One possibility is to start with a coarse time step size depending on the blade passing frequency (BPF), e.g. 40 or 50 physical time steps per period. When the solution has become periodic in time after some iterations the time step size should be refined to resolve the phenomena which are to be analyzed. In most cases a resolution of 100 time steps per period is sufficient. The other approach is to start already with the refined time step but with very little inner iterations (also called pseudo time steps) in order to quickly obtain a periodic response signal. In a second step, the number of inner iterations is increased to guarantee convergence for each time step.

The challenge for both approaches lies in the proper choice of the physical time step. For this study, a criterion from the work of Gourdain and Leboeuf [28] has been used which will be briefly described in the following. The number of physical time steps has to be chosen with particular care to capture all dominant frequencies. The maximum frequency f_{max} that can be computed depends on the mesh resolution and is linked to the dimension of the time step Δt through the Shannon theorem such that

$$f_{max} = \frac{1}{2\Delta t}.$$

(3.34)

This maximum frequency is based on physical considerations in order to correctly consider the rotating structures. In a compressor or turbine stage, tangential modes are induced by the rotor-stator interaction whereas each tangential mode m can be linked to a corresponding wavelength λ_m by

$$m = 2\pi \frac{r}{\lambda_m} = aN_R + bN_s.$$

(3.35)

N_R and N_S represent the number of rotor and stator blades and a and b relative integers respectively. The frequency of the m-order mode is given by

$$f_m = \frac{aN_R\Omega_R}{2\pi}$$

(3.36)

with

$$f_{BPF} = \frac{N_R\Omega_R}{2\pi}$$

(3.37)

being the blade passing frequency. The Shannon theorem enables to relate the smallest circumferential dimension of the mesh l_ϕ to the smallest wavelength λ_{min} that can be computed with this mesh

$$\lambda_{min} = 2l_\phi.$$

(3.38)

This is a minimum condition and due to numerical errors, at least two mesh cells are needed to compute properly a wavelength. Therefore, the greatest mesh dimension of two adjacent cells in circumferential direction determines the minimum wavelength that can be resolved. From equation 3.35, the maximum rotor mode number a_{max} can be estimated for a given stator mode number b and the minimum wavelength λ_{min} obtained from equation 3.38 according to

$$a_{max} = \frac{2\pi \frac{r}{\lambda_{min}} - bN_S}{N_R}.$$ (3.39)

If only temporal modes of the rotor are considered b can be set to zero. Temporal modes of the stator can e.g. be neglected if the focus lies on phenomena resulting from the rotor wake. Then equation 3.40 simplifies to

$$a_{max} = \frac{2\pi \frac{r}{\lambda_{min}}}{N_R}.$$ (3.40)

and the maximum frequency to be resolved with the generated mesh can be calculated by

$$f_{max} = \frac{a_{max} N_R \Omega_R}{2\pi} = a_{max} f_{BPF}.$$ (3.41)

Finally, this leads to an estimation for the physical time step of

$$\Delta t = \frac{2}{f_{max}}.$$ (3.42)

3.3.4 Models for Unsteady Computation

Several approaches for the computation of transient flows in turbomachinery are available. Their common goal is to reduce the computational domain to as little passages as possible and exploit periodicity at the boundaries to accelerate the computation. Unsteady computations in turbomachinery can be divided into two categories, namely time domain and frequency domain methods. Currently the most common time domain method is the transient method or domain scaling method which allows the specification of expressions for unsteady boundary conditions and the use of a sliding mesh at the rotor-stator interface. In this work, the domain scaling method has been employed, which is why this section will be limited to this technique. For further information regarding the other approaches, please refer to [71] and [10].

The Domain Scaling Method

The domain scaling method is based on the constraint that the pitch distance must be identical on both sides of the interface. The major advantage of such configurations is that the flow remains periodic in space with a period equal to the pitch distance. Thus, this method does not impose any time periodicity in the boundary condition and permits to resolve all time frequencies (depending on the mesh resolution) with an error that is proportional to the scaling coefficient.

The domain scaling method considers respectively K_R and K_S blade passages on the rotor and the stator sides with the condition

$$K_R P_R = K_S P_R$$ (3.43)

to be satisfied. Here, P_R and P_S are the pitches of the two adjacent blade rows. In almost all turbomachinery applications the number of rotor and stator blades differ which requires one of the following decisions:

- Modeling a higher number of blade passages. Quite often, this is not possible either because the ratio of rotor to stator blades has been chosen without a common factor to avoid mechanical excitation. In this case, it would be necessary to model the entire machine which is normally not affordable in terms of CPU resources. However, this choice would permit to avoid the introduction of an error due to the geometry scaling.

- Scaling the geometry. The easiest way to obtain the required scaling is to generate a mesh with a blade number different from the real one. However, simply changing the blade number will modify the chord-to-pitch ratio and, hence, may have a major impact on the blade load and the entire flow behavior in the corresponding blade row. The committed error will be rather large. Therefore, a combined scaling should be considered to decrease the error, in which the blade length and thickness are modified accordingly to the azimuthal modification in order to maintain the original solidity.

In some cases, the most suitable treatment might be a combination of the above approaches to find the optimal ratio between memory increase and geometry scaling. A detailed study on this issue is demonstrated with the 1.5 stage Aachen Turbine in [10].

Due to the same pitch distances on both sides of the interface the flow is time periodic in the circumferential direction over the distance KP at any time

$$\phi_S(r,\theta,z,t) = \phi_R(r,\theta - mKP,t) \tag{3.44}$$

and the rotor-stator-interface is implemented as a non-matching connection. The parameter m is selected in a way that the location of the point $(\theta - mKP)$ lies inside the computational domain of the stator. It is beneficial to resolve the unsteady governing equations in the relative frame of reference to avoid the need for grid rotation at each physical time step. Then, only the interpolation data structure needs to be updated in function of the relative position of each row.

The drawback of the domain scaling approach is that the simulation is performed on a modified geometry. Since the amount of scaling may not only affect the capturing of unsteady phenomena but also significantly modify the steady-state flow characteristics and the global machine performance, it is strongly advisable to compare the modified (scaled) geometry to the original geometry in steady state before starting to analyze the configurations in unsteady mode.

3.4 Optimization Method

The design process of modern turbomachinery is a complex and time-consuming task involving many different objectives and constraints. Until the 80s, this was even more challenging since almost the entire design had to be done analytically by the engineer. As a consequence, designers often started from an existing axisymmetric contour and tried to adapt and improve it based on trial and error. However, mostly only a limited amount of prototypes are affordable due to restricted budgets and the request for very short design time schedules. Fortunately, computational power has increased enormously over the last decades helping the designer to define advanced blade geometries, compute the flow field inside the blade channels and the mechanical stresses in the solid parts of the machine.

Although CFD Software is getting more and more accurate, fast and user friendly it normally does not provide algorithms which are able to automatically optimize the performance of a new geometry. Due to the above mentioned restrictions, designers cannot take full advantage of the huge potential and amount of information that is provided by CAD, CFD and structural codes. The optimization problems associated with turbomachinery design often involve many constraints and a large set of parameters which result in objective functions presenting many extrema. It is well known that optimization methods based on gradients techniques are efficient in terms of convergence rate but are not guaranteed to find the global optimum [93]. In contrast, genetic algorithms offer the advantage of enhancing the probability of reaching the global optimum but, unfortunately, may require thousands of iterations [27]. Their coupling with a three-dimensional Navier-Stokes solver is time-consuming and thus difficult within the framework of an industrial design process. The major idea of the optimization approach [11] to be presented in this paper is that the evaluation of the successive designs is performed using an artificial neural network instead of a flow solver permitting to use the genetic algorithm in an efficient manner. The accuracy of the optimization depends on the knowledge of the neural network which is fed by design examples stored in a database. This optimization strategy is used in order to design non-axisymmetric end walls for a compresser design, namely the known Configuration I of the Darmstadt Transonic Compressor [63, 64].

3.4.1 General Principles of the Design Method

The idea of the presented method, of which a flow chart is shown in Figure 3.1, is to accelerate the design of new end wall contours by using the knowledge acquired during the previous designs of similar contours. The evaluation of the successive designs is performed using an artificial neural network instead of a flow solver allowing the genetic algorithm to be used in an efficient manner. The accuracy of the optimization depends on the knowledge of the neural network which is fed by design examples stored in a database. The entire optimization process consists of the following steps:

- Parametrization of the geometry to reduce the geometrical degrees of freedom and creation of a meshing template for the grid generator

- Definition of the free and fixed geometrical parameters, generation of the initial databases based on CFD simulations of arbitrary geometric samples

- Definition of the optimization goals, run of the optimization by means of the artificial neural network, based on a self-learning algorithm and performance check to compare the current sample with regard to the optimization goals

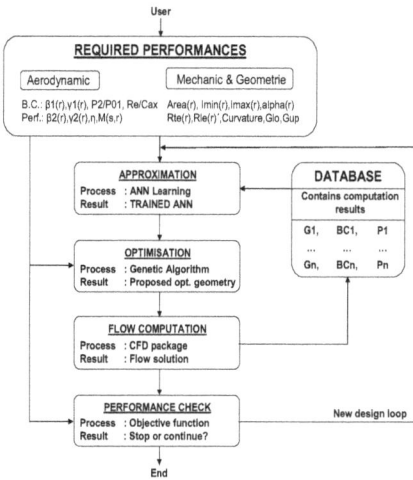

Figure 3.1: End wall design algorithm [72]

The core of the design system is the database containing the results of all Navier-Stokes computations performed during the generation of the initial databases. This database contains three kinds of data for each sample:

- The fluid properties and flow-field boundary conditions used by the Navier-Stokes solver

- The parameters to define each sample's geometry

- Each sample's aerodynamic performance characterized by efficiency, total pressure ratio, the outlet flow angle distributions, mass flow, blade loading and other quantities

After a sufficiently large initial database of arbitrary samples has been generated in an automated way, an iterative procedure is used which is structured as follows:

- A learning process is used to build the neural network based on the examples stored in the initial database. The network contains free parameters that are to be adapted to fit the database samples. A fitting process is performed by back-propagation of the errors. After the mapping of the database samples, the neural network is able to predict the aerodynamic performance of end wall contours that are not part of the database.

- The next step is to find a new design by using the mentioned optimization procedure formed by a genetic algorithm. The performance of this new design is evaluated by means of the trained neural network. To quantify performance, an objective function is used which transfers all user-imposed constraints into a single number. The result of this optimization is a point which is expected to be close to the global optimum of the design space.

- This new geometry is now evaluated by the 3D Navier-Stokes flow solver and added to the database. The comparison of the performance obtained by CFD with the one predicted by the neural network permits to evaluate the accuracy of the network. The new sample's performance is also compared to the design goals. If these have not been achieved another design iteration is started repeating the same process until the optimum geometry has been reached.

The database grows after each iteration leading to improvements of the approximate relation and therefore to a better prediction of the real optimum. The key is the enrichment of the database in the proximity of the optimum design while the database in the less interesting areas of the design space remains relatively coarse. Within the optimization process a sample will be handled as invalid if there is e.g. a failure of the grid generator. The genetic algorithm will then not search for a new possible better design in the proximity of the invalid sample within the next iterations. If the failure is due to an issue such as no available license at the moment of post-processing of a valid sample, then the algorithm will approach this region again after some iterations if the optimum is really located there.

3.4.2 End Wall Parametrization with AutoBlade

The geometry parameterization is a critical element for the success of any shape optimization method. The parametrization should be able to generate a large variety of physically realistic shapes with as few design variables as possible while providing smooth aerodynamic shapes. The used tool is a parametric geometry modeller specifically developed for turbomachinery applications. Designers usually work by generating two-dimensional blade sections which are then stacked along the blade span to obtain the three-dimensional blade geometry. AutoBlade offers a number of design modes for the two dimensional blade sections. A suitable one in the case of axial compressors is to generate a camber curve and to add two additional construction curves for the blade suction and pressure side respectively.

Concerning the end wall parameterization, many approaches can be found throughout literature which rely on shape functions such as sinusoidal curves [7, 29, 35, 54, 79]. Such simple shape function help to conserve the cross passage area and therewidth the capacity of the turbomachinery. However, they also imply an assumed shape of how the resulting design should look like which may lead to an unintended restriction of the design space even though the true optimal contour could differ. More sophisticated shape functions based on Fourier coefficients compensate this drawback to some extent but the correlation between input parameters and the resulting end wall geometry becomes less obvious [73].

To avoid such limitations, a parametric-based technique for defining the contoured end wall shapes has been used in this optimization framework. As tools, the used modeller offers Bezier, B-spline or C-spline curves. To parameterize the end wall perturbation area, cuts based on Bezier curves were built along a virtual streamline. The virtual streamline is aligned with the blade camber curve. Using this method, the cuts may extend outside the blade channel by a user-defined distance. In the Darmstadt Transonic Compressor the parameterized area started upstream of the leading edge and ended downstream of the trailing edge, each by 10% of the blade channel's meridional extension, as shown in Figure 3.2. The blade channel was divided into 5 cuts, each with a distance of 25% of the blade channel width to the next one. A uniform distribution of 4 parameters was selected along the cuts. Since the cuts extended outside the blade channel, the start and the end cuts had to be identical to ensure geometrical continuity. This finally led to 16 free parameters for the optimization which was seen as a good balance between predicted aerodynamic benefit and computational cost. In the circumferential direction, the perturbation law is defined by Bezier curves controlled by height parameters (Figure 3.3). To ensure slope continuity, the additional values h1 and h5 in Figure 3.3 were automatically managed by the algorithm and could not be accessed by the user. The final parameterized patch is a loft surface passing through the perturbed cuts. Slope continuity was also regulated at the intersections between the perturbed area and the upstream and downstream axisymmetric parts of the end wall surfaces. Each control point is limited to one-degree freedom by allowing only movement normal to the end wall. The contour shape can be received simply by the control

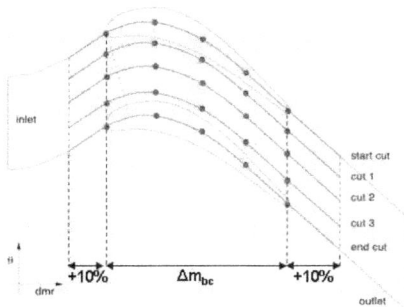

Figure 3.2: Cut definition for method 'Along Channel'

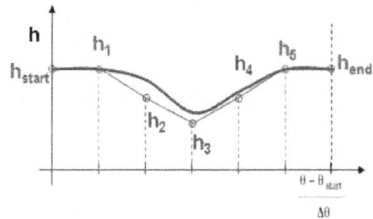

Figure 3.3: Circumferential perturbation law for the blade channel

point values as these values define the deformation displacements from the datum contour. This parameterization was used for all optimizations carried out. For the stator optimization, all parameters were allowed to vary between -5mm and 5mm for both the hub and the shroud end walls. This corresponded to a 7% span with a stator channel height of 70mm. For the subsequent rotor optimization, all parameters were allowed to vary within the same bounds, this time of course only for the hub end wall. In that case, this corresponded to a 6% span with a rotor channel height of 85mm.

As the chosen form of end wall parameterization does not contain periodic functions in the pitch-wise direction, significant variations in throat area and therewith capacity may follow. To keep this effect as small as possible, a throat area correction scheme has been incorporated into the design system by introducing a geometrical constraint. This was done by applying a penalty term on deviations from the original throat area (see chapter 3.4.6). AutoBlade also provides the possibility to analyse the 3D blade shape by computing quantities such as blade thickness, momentum of inertia and curvature at different locations. Afterwards, it is possible to include these quantities in the objective function and discard samples which are invalid due to mechanical constraints.

3.4.3 Grid generation with AutoGrid

AutoGrid is an automatic meshing system for turbomachinery applications developed to ease pre-processing for numerical computations on such configurations. It contains many features to enable the user to create meshes for a large range of gas turbines, fans and compressors applications (e.g. Turbofan, Turboprop, axial or centrifugal, single or multistage). The available topologies lead to block-structured grids of high quality. A big advantage is the possibility to work with meshing templates, i.e. to define mesh properties as grid point distribution, tip clearance etc. and apply them to other configurations by using the template. This is particularly important concerning an optimization due to the geometry modification of each iteration. First of all, a mesh must be generated for the parameterized initial geometry with a grid quality as high as possible to avoid poor grid qualities afterwards being possible to appear due to the desired geometry modifications. In a multistage project, as described here, the rotor and stator are to be meshed separately in two projects since it has only been possible to parametrize and modify one blade row (either rotor or stator) within the FINE/Design 3D environment at the time of the project.

3.4.4 The Approximate Model

The basic principle of the method is to build an approximate model of the original analysis problem. This model can then be used inside an optimization loop instead of the original model. In this way the performance evaluation by the approximate model is not costly because no CFD simulation is needed for this purpose. Thus, the number of performance evaluations executed by the model for the optimization is no longer critical. Among the large number of possible techniques able to construct such a model, an artificial neural network (ANN) has been selected. Even though the initial motivation for the development of the ANN was to establish computer models being able to imitate certain brain functions, ANN can also be considered as a powerful multi-dimensional interpolator. Artificial neural networks are non-linear models that can be trained to map functions with multiple inputs and outputs. The ANN consists of series of layers, each of them being composed of a given number of nodes. The first input layer connects all inputs (i.e. the non-axisymmetric end wall geometry and the constraints) to the network whereas the last output layer produces the outputs (i.e. the aerodynamic performance). All inputs of a layer are connected to all nodes of this layer through weighting factors. As schematically described in Figure 3.4, the summation of all the contributions with a corresponding additional bias value is introduced in a sigmoidal function F where the output of each node is given by

$$a_n(i) = F \left[\sum_{j=1}^{S(n-1)} W_{ij} a_{n-1}(j) + b(i) \right] = F \left[i_n(i) \right] \tag{3.45}$$

The training of the neural network consists of finding the components of the weight matrices that lead to the best reproduction of the examples stored in the database trying to minimize the error that the ANN produces on the database samples.

$$E^l = 0.5 \sum_{j=1}^{nout} (d_j^l - a_j^l)^2 \tag{3.46}$$

Here, d_i and a_i represent the database and the corresponding ANN values of the first output for a given sample I. This optimization problem can be solved by using a gradient method, which is usually referred to as the back-propagation approach. The derivatives of the error regarding the weight factors are expressed which permits to calculate iteratively the weight and bias components modifications to eliminate the error. For a more detailed description of the used artificial neural network, please refer to Pierret et al. [72] and Hildebrandt & Thiel [46].

3.4.5 The Optimization Algorithm

The goal of the optimization is to find the minimum of the objective function using the simplified analysis model. In this context, an essential issue is the robustness of the numerical

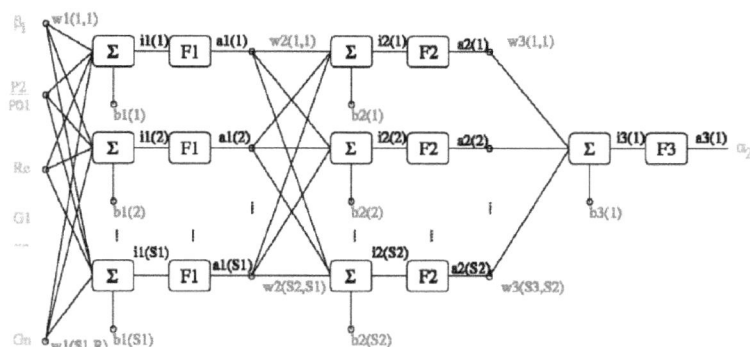

Figure 3.4: Schematic view of an ANN [46]

optimization algorithm. The choice of the optimization algorithm is mainly based on the following two considerations:

- Many local optima may exist in the design space and therefore a global optimization technique is required.

- The evaluation of the performance using the approximate model is very fast (a few ms). Consequently, the number of needed function evaluations is now of far less importance than if a detailed CFD computation was needed at each step.

Based on the first consideration, the straightforward application of numerical optimization techniques that rely on derivatives is questionable because of its local properties. In contrast, stochastic techniques such as the genetic algorithm or simulated annealing are global optimization methods that do not get stuck in a local minimum and thus offer an alternative to conventional gradient methods which, however, have the advantage of a fast function evaluation. Genetic algorithms were designed by Holland in the 70s and improved by Goldberg in the 80s [27]. A genetic algorithm is summarised as follows. An initial population is generated by randomly selecting individuals in the whole design space. Then, pairs of individuals are selected from this population based on their objective function values. The performance of an individual is evaluated by its fitness. After that, each pair of individuals undergoes a reproduction mechanism to generate a new population in such way that fitter individuals will spread their genes with higher probability. The children replace their parents. As this proceeds, inferior traits in the pool die out due to their lack of reproduction. At the same time, strong traits tend to combine with other strong traits to produce children who perform even better.

3.4.6 The Objective Function

Among the design objectives of a detailed aerodynamic shape optimization, efficiency is only one of the many considerations. A good design must also satisfy the mechanical, manufacturing and aerodynamic constraints. In the presented case, this may be the outlet flow angle profile, the throat area due to the contouring's impact on the cross section, and the pressure loss coefficient. The optimization problem can be seen as the minimization of an objective function in function of several variables (geometrical parameters) subject to several constraints (mechanical, manufacturing and aerodynamic constraints), the objective function and the constraints being non-linear. The general approach for solving this problem is to transform the original constrained minimization problem into an unconstrained one by converting the constraints into penalty terms that will increase when the constraints are violated. Summing up all penalty terms and the original objectives then creates a pseudo-objective function

$$OF = \sum_{Penalties} w * P. \tag{3.47}$$

Here, w are weighting factors that the user has to associate to each penalty. One can notice that the difference between the actual value V and the required one V_{req} is divided by a reference value V_{ref}, in order to provide all terms with a similar order of magnitude. Depending on the type of constraint, the penalties will have the following formulations:

- Upper bound: if $V > V_{req}$ then

$$P = \left(\frac{V - V_{req}}{V_{ref}} \right)^2, \qquad \text{else P=0} \tag{3.48}$$

- Lower bound: if $V < V_{req}$ then

$$P = \left(\frac{V_{req} - V}{V_{ref}} \right)^2, \qquad \text{else P=0} \tag{3.49}$$

- Equality constraint:

$$P = \left(\frac{|V - V_{req}|}{V_{ref}} \right)^2 \tag{3.50}$$

The optimization technique that is adopted in FINE/Design3D can be considered as single objective since all objectives and constraints are put together into one single function. Weighting functions are associated to the different constraints allowing the user to reflect the levels of priority into the optimization. Since NUMECA does not offer the possibility of a multi-objective optimization to create pareto fronts, yet, the optimum geometry depends on how the user chooses the weightings and which terms he includes in the objective function. Therefore, different weightings will lead to different optimum geometries.

4 Stator Optimization

In this chapter, the first test case for the Configuration I of the Darmstadt Transonic Compressor is presented. The challenge is to find the optimal end wall shape that best satisfies the objectives of reducing secondary losses and increasing efficiency. For this design exercise, the optimization system of NUMECA (see chapter 3.4) was used to find an adequate end wall shape. It was sequentially applied to the stator end walls of Configuration I of the Darmstadt Transonic Compressor. The hub end wall was firstly optimized at design conditions and the casing afterwards at off-design conditions near stall. A general outline of this configuration is given in Table 4.1. Table 4.2 shows the main aerodynamic design parameters for the stator blade. A further description of Configuration I can be found in Müller [63, 64]. As the blading of the stator is almost purely 2D, the results of the generated non-axisymmetric end wall geometry have to be discussed in terms of their potential in state-of-the-art compressors with 3D bladings.

4.1 Numerical Procedure

In this section, an overview of the computational methodology is given. All calculations throughout the entire optimization process were run as steady single passage RANS simulations using the EURANUS solver of the CFD Package NUMECA/FineTurbo. The code solves the the RANS equation on structured multi-block arbitrary grids. The solver is based on a cell-centered approach associated with a second-order finite volume discretization and an explicit Runge-Kutta procedure. Acceleration techniques such as multi-grid cycles and parallelization were also incorporated to improve convergence. As already mentioned in section 3.2 the one-equation model of Spalart-Allmaras was used for turbulence modeling. Although the objective was to optimize the stator, the rotor was included in the CFD simulations. Accordingly, the stator row was included in the CFD simulations of the rotor design study which will be discussed in chapter 5. Both rotor and stator were meshed with a 4HO topology (skin block around blade uses a O-topology, the inlet, outlet, upper and lower blocks use an H-topology). The computa-

Corrected mass flow rate (Design point)	16.0 kg/s
Total pressure ratio (Design point)	1.5
Design shaft speed	20,000rpm
Number of blades - rotor/stator	16/29
Tip diameter	0.38 m
Inlet relative Mach number at tip	1.35
Inlet relative Mach number at hub	0.7
Hub-to-tip-ratio rotor	0.51
Hub-to-tip-ratio stator	0.63

Table 4.1: Outline of Configuration I

Stator design parameter	Hub	Mid	Tip
Turning [°]	43.83	35.86	35.71
ΔV_{φ} [m/s]	159.69	124.54	112.95
Chord length [m]	0.049	0.052	0.055
Stator solidity (chord/pitch)	1.95	1.52	1.34
Diffusion factor	0.41	0.37	0.37
Aspect Ratio	0.7	0.74	0.78

Table 4.2: Aerodynamic blade design parameters

tional grids of both rows are shown in the blade-to-blade view in Figure 4.1 and 4.2. The rotor grid consists of around 60 nodes in the blade-to-blade direction, 135 nodes in the streamwise direction including an additional H-block upstream a defined z-constant line and 141 nodes in spanwise direction. Concerning the distribution in the spanwise direction, a constraint of 80% mid-flow cells was set to ensure a high resolution in the end wall region, being the main section of interest. The rotor tip clearance is represented by 33 nodes in blade-to-blade direction, 98 nodes in streamwise direction and 21 nodes in spanwise direction. The grid of the stator contains 51 nodes in the blade-to-blade direction, 101 nodes in the streamwise direction and 117 nodes in spanwise direction. Again, a constraint of 80% mid-flow cells was set to ensure a high resolution in the end wall regions. For all simulations, the y^+-value was below 5.

Constant values of total pressure and total temperature were applied as inlet boundary conditions, the flow direction being in axial direction. As the outlet boundary condition for the optimization, the mass flow was set to the desired value. The intention was to remain at the same operating point during the optimization process. For the rotor-stator-interface, the mixing plane approach was used which is a standard type for industrial applications. It only requires one passage per blade row. The flow in both domains is computed as a steady state problem. All flow variables at the interface are averaged in circumferential direction for the rotor outlet as well for the stator inlet. Thus, transient effects due to the interaction of rotor blades and stator vanes are not resolved. An option with non-reflecting boundary conditions was used to avoid non-physical reflections between the two blade rows.

However, the speed lines were calculated by successively increasing the static back pressure. When approaching stall, the exit boundary condition was switched to the mass flow constraint again because the numerical solution tended to diverge at high back pressures. With the mass

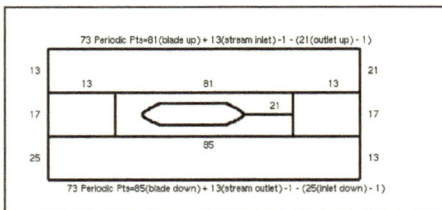

Figure 4.1: Grid point distribution for rotor in blade-to-blade view

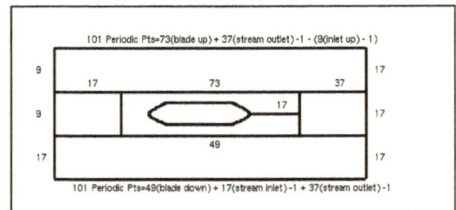

Figure 4.2: Grid point distribution for stator in blade-to-blade view

flow condition applied it was possible to calculate some more operating points closer to stall. The numerical solution in Figure 4.3, referred to as the numerical stability limit, corresponds to the minimum possible mass flow which permitted a fully converged steady solution.

4.2 Settings of the End Wall Optimizations

Both end wall shapes were designed through a single-point optimization at a design speed of 20,000rpm. Regarding the hub end wall, the optimization was carried out at a reduced mass flow rate of 16.1kg/s where the compressor has a peak-efficiency of 85.37%. The casing end wall was optimized at off-design conditions at a reduced mass flow rate of 15.25kg/s. Both optimizations were run at constant mass flow rates to remain at the same operating points throughout the whole procedure. To justify this approach we shall forestall that the stator of the investigated Configuration I shows huge areas of separation, especially in the hub region. The intention was to isolate the effects that might result from the hub optimization and moreover to derive whether these improvements may have any negative impact on the casing region of the stator or not at the same time. This in turn could be of major importance when increasing the blade load since Configuration I of the Darmstadt Compressor is rotor-tipcritical for the onset of stall

The objective function to be minimized of the hub design process was defined by a combination of five terms: isentropic stage efficiency, stator pressure loss coefficient, stator exit whirl angle (mass-weighted average), SKE at the stator exit plane and the stator throat area to avoid effects that could occur by purely changing the cross section without end wall profiling. Moreover, it should be ensured to maintain the reaction of the stage and thus keeping the rotor on the same part of its characteristics. Although minimizing SKE is a precedence for turbine optimization, it was also included in the compressor optimization.

Usually, efficiency is calculated with the total pressure and total temperature considering the entire velocity vector, which consists of a primary and a secondary component. Therefore, secondary flow is included in the efficiency. A flow that contains high secondary components could still have a high stage efficiency when using total values for calculating this parameter. The shear angle between secondary and main flows leads to an increased dissipation of energy resulting in pressure losses. SKE can therefore be seen as the losses of the following downstream stage. The whirl angle was included as it was not clear to what extent end wall profiling would lead to under or overturning of the flow. Equation 5.1 represents the objective function.

$$OF = w_\eta P_\eta + w_{CP} P_{CP} + w_\alpha P_\alpha + w_{SKE} P_{SKE} + w_A P_A \qquad (4.1)$$

Here, w are the weighting factors of the linear combination of penalties, a penalty term given by

$$P = \left(\frac{|V_{req} - V|}{V_{ref}} \right)^2 . \qquad (4.2)$$

V_{ref} makes the expression dimensionless. In this context, the optimum geometry depends on how the user chooses the weightings and which terms he includes in the objective function. Therefore, different weightings will lead to different optimum geometries, which is the critical part of the optimization. The highest priority was given to efficiency as it was the most important factor (~44%). Half the influence was set on the cross section (~22%) since this constraint was not expected to be connected to the flow parameters. The remaining percentage was equally distributed among the three other terms. Target values, weighting factors and the resulting influence in percent of the penalties are summarized in Table 4.3. For the design process of the shroud end walls, the objective function was reduced and the term of the SKE no longer considered. As it can be seen in further discussion, an increase in efficiency went along with a large reduction in SKE in the hub optimization. However, this statement is not universally valid for compressors and will only be applied to the analyzed stator geometry. Evaluation of the SKE was omitted due to its time-consuming external routine. The penalty for isentropic efficiency was expected to cover both aspects. Also, the weighting factors were changed to have a higher influence on the efficiency term and to give less priority to the cross section as this had not changed significantly during the hub optimization process. An overview is given in Table 4.4.

	η [%]	CP_{Stator} [-]	α [°]	SKE [m^2/s^2]	Throat area [m^2]
Required value	100	0	0	0	0.001971
Weighting factor	10	8	0.03	0.0002	25000
Weighting in percent $w_i P_i / OF$ [%]	43.6	10.5	13.2	11.1	21.6

Table 4.3: Parameters for the objective function - Hub Optimization

	η [%]	CP_{Stator} [-]	α [°]	Throat area [m^2]
Required value	100	0	0	0.001971
Weighting factor	22	8	0.03	800
Weighting in percent $w_i P_i / OF$ [%]	70.25	6.18	17.45	6.12

Table 4.4: Parameters for the objective function - Shroud optimization

4.3 Optimization Results - Overview

This section gives an overview of the results and illustrates the observed phenomena. Attempts at explaining the improvements in efficiency and overall performance are given separately for both optimization steps in the following subsections. Figure 4.3 shows the efficiency characteristics of the datum axisymmetric design, the first optimized design with contoured hub only (called **Optimization A** in the following) and the final geometry with both end walls contoured (called **Optimization B** in the following). In all cases, the isentropic efficiency was calculated with the mass-averaged values of total temperature and total pressure at rotor inlet and stator outlet respectively. In addition, the experimental results of the datum design are displayed by the black symbols. Since this is a numerical study, no experimental data of the contoured designs are available.

Optimization A led to an astonishing increase in efficiency of 1.8% at design conditions. Comparing the red and blue lines, one can note that the hub optimization improved the entire speed line even though the benefits are less near stall. With 14.6kg/s, the non-axisymmetric hub end wall design, unfortunately, features its numerical surge limit (last converged solution) at a higher flow rate than the datum one.

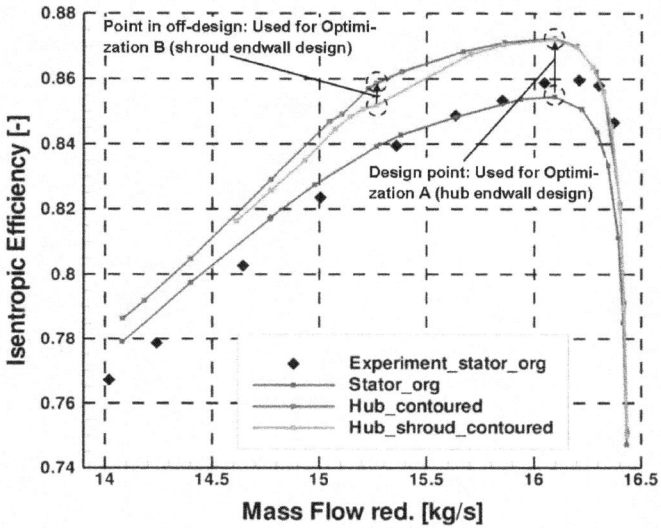

Figure 4.3: Isentropic efficiency vs. mass flow characteristics of the different design steps [76]

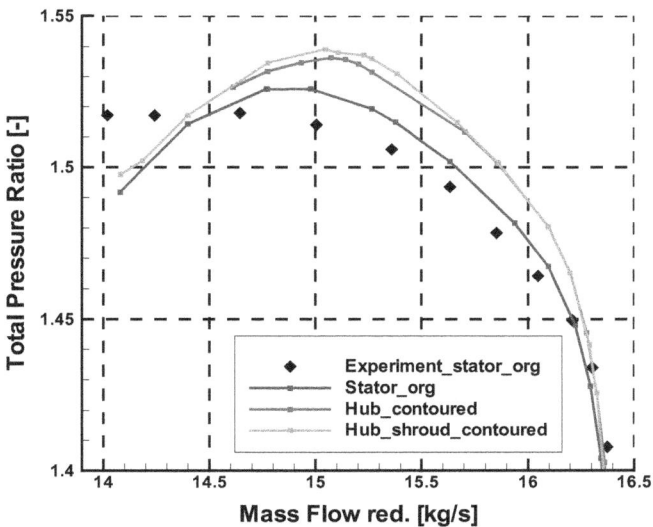

Figure 4.4: Total pressure ratio vs. mass flow characteristics of the different design steps [76]

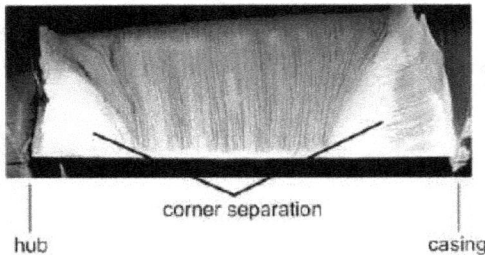

corner separation

hub   casing

Figure 4.5: Oil-streak pattern on the suction side of analyzed stator, taken from Hergt et al. [44]

With the experimental setup, the last stable point was measured at a reduced mass flow rate of 14.0kg/s which could almost be reached in the numerical analysis of the datum geometry. At approaching stall conditions the usually increasing level of unsteadiness (rotating stall, local separations) tends to render steady CFD simulations more difficult and prone to errors. The expression 'numerical stability limit' will therefore be used instead of 'surge limit' in the following. Optimization B resulted in a further remarkable improvement for the off-design operational range. In comparison to the hub contoured only design, an additional gain in efficiency of 0.67% could be reached in the conditions used for this subsequent design of the casing end wall. The numerical stability limit of the datum design could be reached again. It seems to be sensible that the optimizations have been carried out at constant mass flow rates, trying to keep the throat area constant. The characteristic line was therefore not shifted along the x-axis, which would have made a comparison quite difficult.

Figure 4.20 shows the total pressure vs. mass flow characteristics for the different designs. The curves get steeper with the optimized geometries. The optimized hub features the lowest stall line. In contrast, the sequentially optimized design shows a higher stall line than the original geometry as a higher total pressure ratio is reached at the last stable point featuring the same mass flow for both designs.

Figures 4.6 to 4.11 give an initial idea where the changes in performance come from. They all show areas of reverse flow, indicated by iso-surfaces with $V_{ax}<0$ m/s. Figures 4.6 to 4.8 represent the different optimization levels at design conditions and Figures 4.9 to 4.11 near stall. Comparing Figure 4.6 to the experimental observations in Figure 4.5, the CFD seems to capture the separation area at the hub quite well, but, showing some weaknesses in predicting the extent of separation at the casing. In Figure 4.7 the hub-corner stall of the original geometry (Figure 4.6) could be completely avoided by using Optimization A, i.e. the application of non-axisymmetric end walls to the stator hub. According to the stator design parameters in Table 4.2, in particular the diffusion factor, the separation occurs due to poor design and not high blade loading. The prevention of this loss source explains the enormous jump in efficiency over a wide operating range. However, in Figure 4.7 it also becomes apparent that the region of reversed flow in the tip region increased remarkably at the same time, now also indicating a recirculation instead of only a downward deflection in radial direction. This behavior is even more distinct near stall, as illustrated by the comparison between Figures 4.9 and 4.10. Concerning the datum design, the separation area is associated with a blockage, decreasing the effective cross section of the flow. At constant mass flow the existing blockage of the hub separation leads to a local acceleration, delaying the shroud separation. After reducing hub-corner stall by using Optimization A, the same 3D separation phenomenon (tip-corner stall) appears at the casing as before at the hub surface. This can also be considered as the reason why the two corresponding characteristic lines get closer at around 14.6kg/s. Nonetheless, there is still a small gain in efficiency due to the smaller dimension of the tip-corner stall compared to the hub-corner separation.

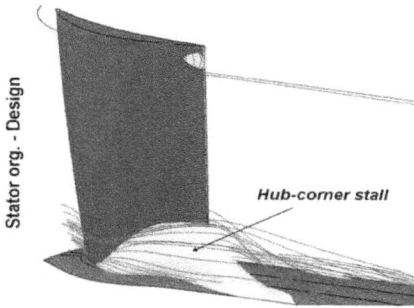

Figure 4.6: Reverse flow area ($V_{ax} < 0$ m/s) for **original** stator geometry at **design conditions** (16.1kg/s) from [76]

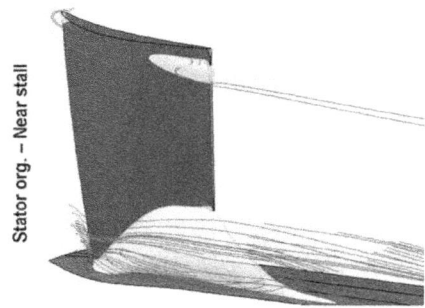

Figure 4.9: Reverse flow area ($V_{ax} < 0$ m/s) for **original** stator geometry at off-design **near stall** (15.25kg/s) from [76]

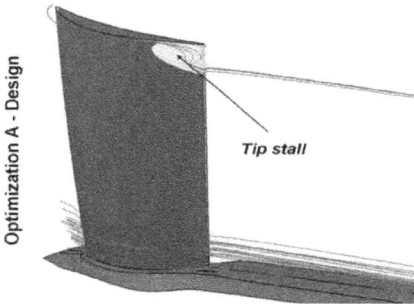

Figure 4.7: Reverse flow area ($V_{ax} < 0$ m/s) for stator geometry with **optimized hub** contour at **design conditions** (16.1kg/s) from [76]

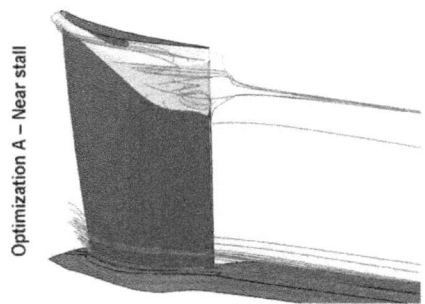

Figure 4.10: Reverse flow area ($V_{ax} < 0$ m/s) for stator geometry with **optimized hub** contour **near stall** (15.25kg/s) from [76]

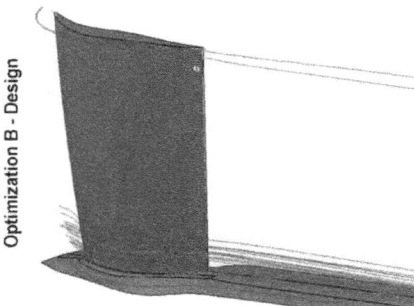

Figure 4.8: Reverse flow area ($V_{ax} < 0$ m/s) for stator geometry with **optimized hub and shroud** contours at **design conditions** (16.1kg/s) from [76]

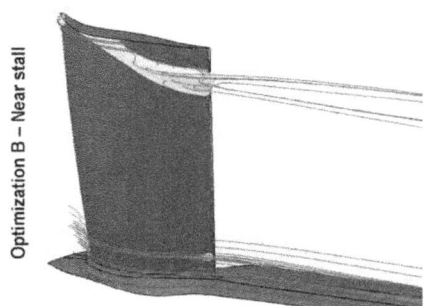

Figure 4.11: Reverse flow area ($V_{ax} < 0$ m/s) for stator geometry with **optimized hub and shroud** contours **near stall** (15.25kg/s) from [76]

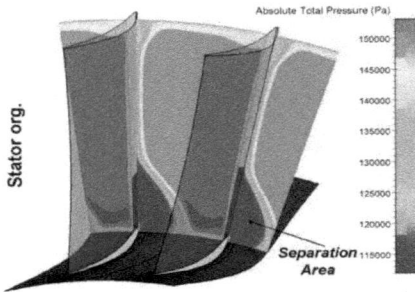

Figure 4.12: Losses due to separation indicated by low total pressure near stall - original stator geometry from [76]

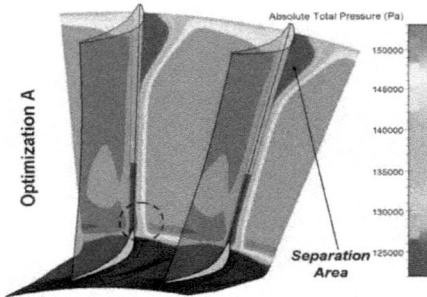

Figure 4.13: Losses due to separation indicated by low total pressure near stall - stator geometry with optimized hub from [76]

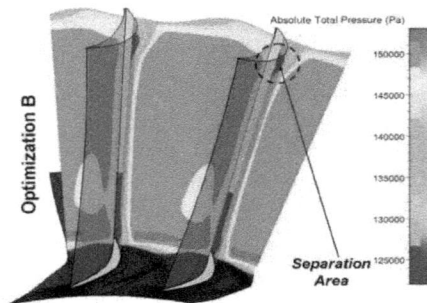

Figure 4.14: Losses due to separation indicated by low total pressure near stall - stator geometry with optimized hub and shroud from [76]

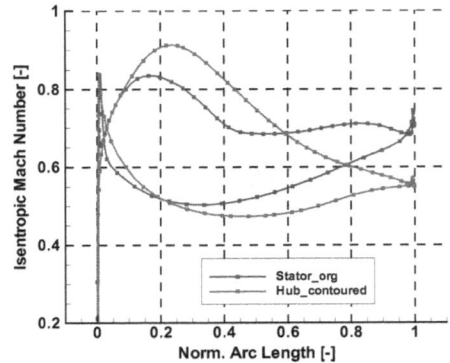

Figure 4.15: Isentropic Mach number distribution at 10% channel height - peak efficiency

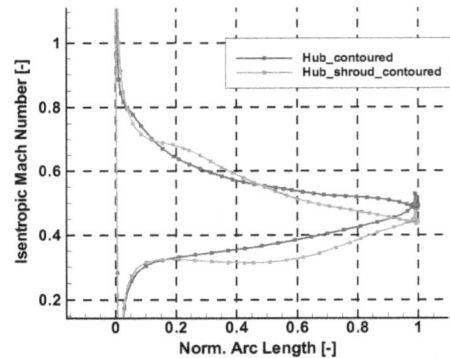

Figure 4.16: Isentropic Mach number distribution at 95% channel height - near stall

The limit on the numerical surge line is no longer set by stator hub-corner stall but determined by the separation occurring close to the stator casing. This is important as Configuration I of the Darmstadt Compressor is rotor-tip-critical for onset of stall. Separation in the tip region of the stator is expected to have a negative influence on locally throttling the rotor in the tip region (see next section). Due to the experience gained throughout Optimization A, the operating point with the pronounced tip separation area near stall was used for Optimization B of the shroud end wall.

4 Stator Optimization

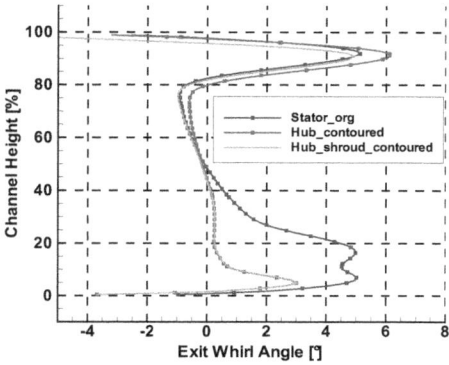

Figure 4.17: Radial distribution of exit whirl angle for setups at design conditions from [76]

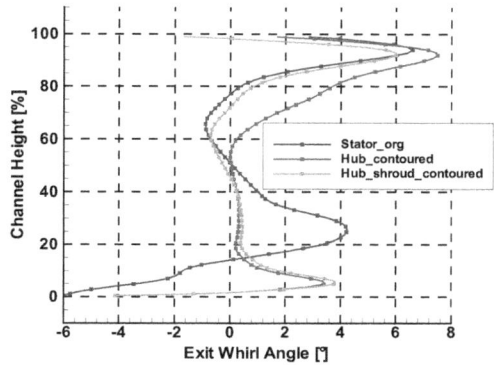

Figure 4.18: Radial distribution of exit whirl angle for setups near stall from [76]

Figure 4.11 displays the area of reverse flow near stall after the second optimization, showing a strong reduction in the separation area in the tip region. This led to a further large improvement in the compressor characteristics at off-design conditions. Even for the design point, an additional small rise in efficiency could be achieved due to less reverse flow (Figure 4.8). This additional benefit of 0.03% in efficiency was not really expected as the contour was designed for different operating conditions.

Figures 4.12 to 4.14 show the total pressure distribution at approximately 75% axial chord length for the different set-ups near stall. The regions of low total pressure once again indicate the extension of the separation bubbles into the blade channel. In Figure 4.12 it is noticeable that hub-corner stall covers about one third of the blade span and almost half of the blade channel, which is in good accordance with Figure 4.6. A comparison between Figures 4.13 and 4.14 shows that the separation could be suppressed in the upper part of the blade by using Optimization B and that it only covers a smaller radial area. Both Figures 4.13 and 4.14 show that the loss cores, indicated by the black dashed circles, have migrated off the end wall onto the suction surface of the airfoil after the corresponding optimization.

Figure 4.15 and Figure 4.16 show the isentropic Mach number distribution at 10% span at design conditions and 95% span near stall respectively. The isentropic Mach number is the Mach number that would appear without any losses in the flow and can be considered as a representation of the blade load. It is calculated according to

$$M_{is} = \sqrt{\left(\left(\frac{p_{ref}}{p_{stat}}\right)^{\frac{\gamma-1}{\gamma}} - 1\right)\frac{2}{\gamma - 1}} \qquad (4.3)$$

where p_{stat} is the static pressure on either the pressure or suction surface and p_{ref} denotes a reference pressure in the free stream. As a reference value in each case, the pitch-wise averaged total pressure at the corresponding relative blade height at the stator inlet was taken. The isentropic Mach number distribution in Figure 4.15 confirms the suppression of the hub-corner stall and the beneficial load changing in terms of making the profile work again. In the case of the original axisymmetric geometry, the flow on the suction side is only decelerated until 40% of the chord length after its maximum velocity. The further horizontal progress is associated with a blockage decreasing the effective cross section for the flow. A further deceleration of the fluid is not possible on the suction side. On the other hand, this diminution in cross section forces the fluid to move to the pressure side increasing its velocity. As a consequence, this yields a recurrent rise of the isentropic Mach number at the rear part of the pressure side. The contoured hub design was found to accelerate the flow more strongly on the suction side resulting in a higher blade loading in the front part of the blade and thus in a higher transverse static pressure gradient. This is not against the objective to reduce or redistribute blade loading by end wall profiling since separation is due to a poor design and not high blade loading. At first sight one would suppose that this disagrees with the reduction of secondary losses. As we will see later in this chapter, this is not necessarily the case. Since the separation involving blockage has been reduced the flow is further decelerated in the back part of the suction side leading to an unloading of the blade in that area compared to the initial case. The contoured end wall design has resulted in a strongly improved static pressure recovery.

Figure 4.16 shows the typical widely open Mach number pattern for a turbomachinery blade experiencing a highly positive incidence. The contoured hub geometry (geometry A) has to deal with a separation from the leading edge. In contrast, the completely contoured geometry (geometry B) decelerates the flow more strongly on the suction surface until around 20% chord length where blade stall starts (note the change in slope). However the blade stall of geometry B occupies much less volume than the tip-corner stall of geometry A further downstream. This leads to a much more favorable suction side diffusion for geometry B in the rear part of the channel whereas geometry A hardly provides any further deceleration along the last 40% chord length. Again, the higher blockage of geometry B in the casing region leads to a decrease in effective cross section which forces the fluid to move to the pressure side increasing its velocity.

In Figures 4.17 and 4.18 the calculated radial distributions of the stator exit whirl angle for all setups at design condition and near stall are compared. The application of the non-axisymmetric hub contour reduced the underturning of the flow enormously. The underturning had been caused by the blockage of the hub-corner stall. At design conditions, it was decreased by 2° near the hub and up to 4° at 20% span. This also resulted in a higher overturning in the hub boundary layer (outflow intended to be axial). This stronger deflection, which is also be observed in turbine applications, is not obvious since the purpose of end wall contouring is to influence the static pressure gradient and consequently reduce the cross flow. The profiled hub end wall also influenced the fluid in the upper 20% of the channel (Figure 4.17). A higher underturning in the tip region was observed, which matches the observation of increasing separation and blockage, as shown in Figure 4.7. Near stall (Figure 4.18) underturning in the tip region for Optimization A even starts at 60% channel height. The deviations from the datum design are also generally bigger than at design conditions. At the same time the exit whirl angle characteristics of the datum design near hub shifted down to smaller radial positions. The strong overturning of the original geometry below 15% blade span was reduced as well as the radial extent of underturning, although the peak values of underturning were not decreased as much as in design conditions. After the additional shroud design within Optimization B the distribution near hub agreed with the one after Optimization A for design conditions. The level of the datum design for underturning near shroud could be reached again. At off-design conditions the additional shroud contour led to a qualitatively similar profile close to the hub whereas the

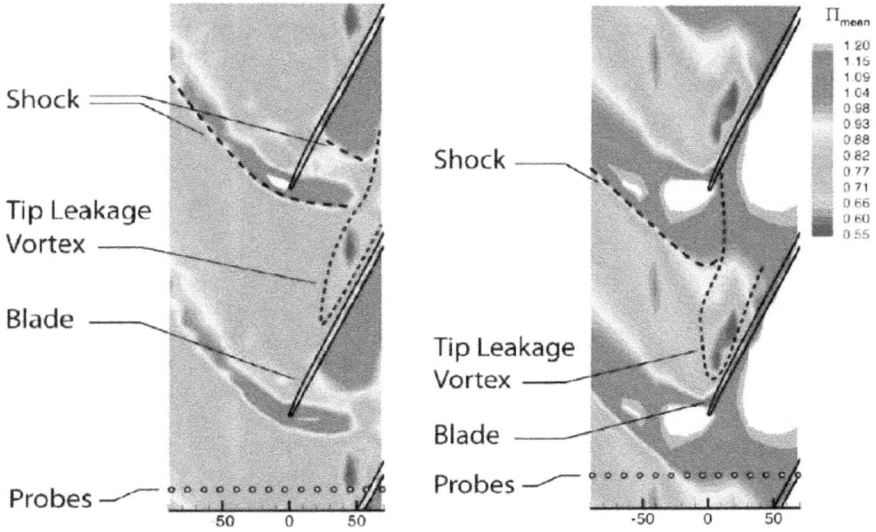

Figure 4.19: Mean static pressure distribution at casing at operating point (left) and near stall (right), experimental data taken from Biela et al. [4]

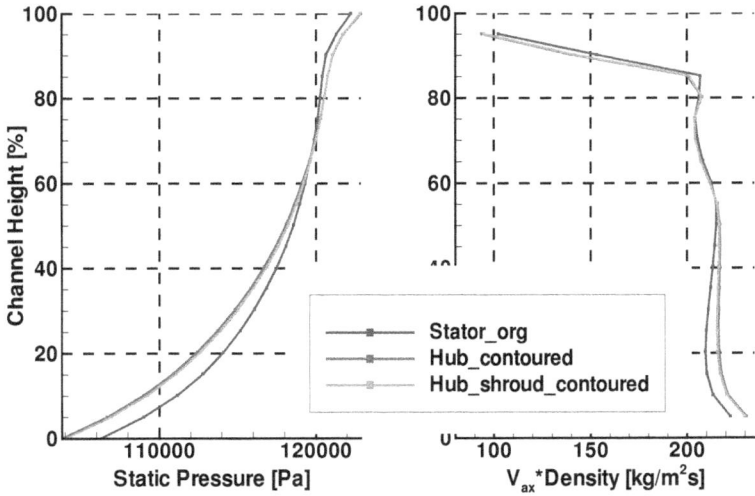

Figure 4.20: Rotor discharge profiles at a mass flow rate of 15.1kg/s - static pressure (left) and mass flow distribution (right) from [76]

underturning came out slightly stronger than with the hub contour only. This is in agreement with Figure 4.11, where a small area of reverse flow is visible close to the hub. This was caused by a blockage, forcing the fluid to move around it. From 50% span onwards the original distribution of the datum design could be approached again. In the upper 10%, underturning could even be reduced, leading to higher overturning in the boundary layer.

4.3.1 Impact on the Rotor

In the previous sections, Configuration I of the Darmstadt Compressor was stated to be rotor-tip-critical for the limit in surge margin. Figure 4.19 shows the static wall pressure distribution at peak-efficiency and near stall. At peak-efficiency, the passage shock is attached at the leading edge and a second shock is visible in the channel. The tip leakage vortex trajectory, which is indicated by low static pressure, reaches deep into the blade channel. When the back pressure is increased towards stall, only one shock is visible and is detached from the blade. The size of the vortex increases and its trajectory does not reach into the channel any more, but rather points closer to the leading edge of the adjacent blade. When the back pressure is further increased, forward spillage of the tip clearance flow can be observed which leads to the onset of rotating stall [33]. In this context, casing treatments have increased the stall margin [64]. To investigate the effect of the profiled stator end walls on this mechanism, the rotor discharge profiles for the static pressure and the mass flow distribution have been plotted in Figure 4.20. The operating point of Optimization B has been selected to compare all three designs (the discharge profiles at design conditions hardly differ from each other). After removing the blockage of the hub separation (Optimization A), the static pressure decreases from the hub up to 50% span. This area is dethrottled and the mass flow increases accordingly. In the tip region one can observe the exact opposite. The new tip separation leads to the expected local throttling, involving a higher static pressure and a lower mass flow rate at the rotor exit compared to the datum design.

Considering what has been mentioned about rotor-tip-criticality, the higher static pressure due to the separation is supposed to be the reason why numerical stability is reached at a higher mass flow rate after the hub optimization. The shroud optimization resulted in a reduction of the tip stall. The smaller blockage leads to a decrease in the static pressure again, even though this is quite small compared to Optimization A. Accordingly, the mass flow rate increases slightly. The tip region of the rotor is therefore dethrottled, although not to the original level. However, this small change seems to be enough to enable the CFD to reach the datum stability limit again. It is imaginable that optimizing the shroud end wall at an operating point even closer to stall could lead to a further dethrottling of the rotor tip region and therefore influence stage stability positively.

4.3.2 Analysis of Optimization A - Hub Design

In Figure 4.21a the end wall shape is shown as a contour of surface distortion relative to the datum axisymmetric end wall. The contour plot shows a strong depression of about -3mm (4.3% span) next to the pressure side and a raised region in the rear part of the blade channel. A smaller rise is observed in the middle front part of the passage and a depression on the suction side. The lowering next to the pressure side is particularly against the expectations and the

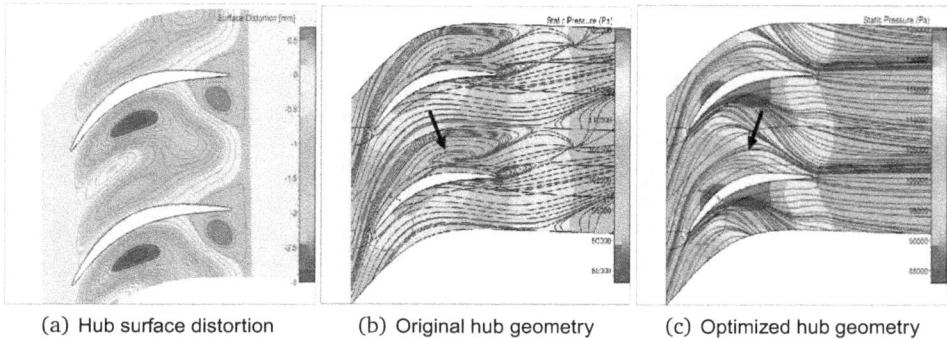

(a) Hub surface distortion   (b) Original hub geometry   (c) Optimized hub geometry

Figure 4.21: End wall height contour of optimized hub (a); Comparison of static pressure and streaklines on hub surface (b, c) at design conditions from [76]

experience of turbine researchers as this is assumed to lead to an even higher static pressure on the pressure side, resulting in turn in a stronger transverse static pressure gradient. To help understand the contradictions present, contours of static pressure together with streaklines on the hub surface are shown in Figure 4.21b and 4.21c. The streaklines were assured to begin at the same starting points for all particle paths. The particle paths highlight the suction side separation line of the original geometry (marked red in Figure 4.21b) This line is driven across the passage towards the pressure side since the flow has to move around the separation area which acts as an obstacle by blocking the cross section.

It also clarifies the extent of hub-corner stall in the original stator. Comparing the static pressure contours, we can see that the static pressure on the pressure side was remarkably increased by end wall contouring which corresponds to the depression in 4.21a. On the other hand, one can also note a lower static pressure at the early suction side surface although the contour has been decreased in this area as well. This is due to the prevention of the separation enabling the profile to accelerate the fluid further. The cause for all these observations is the changed pressure distribution. Although the static pressure gradient is now higher in the circumferential sense, its effective direction, indicated by the black arrows, changed considerably. Regarding the contoured case, the static pressure gradient is almost parallel to the streaklines in the throat area region and thus the effective force pushing the cross flow from pressure to suction side is drastically reduced. At the leading edge the streaklines are further away from the suction side (marked with red dashed circle) which opposes the usual direction of the secondary flow. In contrast, the static pressure gradient is almost perpendicular to the streaklines in the rear part of the channel, enhancing the cross flow from pressure to suction side. This leads to the observed overturning in Figure 4.17b. This effect of encouraging flow into the end wall region causes the loss core to migrate from the hub end wall to the airfoil, as seen in Figure 4.13. This migration of low momentum fluid is very important in terms of delaying or even avoiding corner stall [39] and is similar to the effect of sweep and lean applied to 3D airfoils [86]. Returning to the datum design in Figure 4.12, this low energy fluid remains on the end wall and causes the observed corner stall. Figures 4.22 and 4.23 show the SKE distribution of the original and the contoured hub geometry one axial chord length behind the stator exit. In the contoured case,

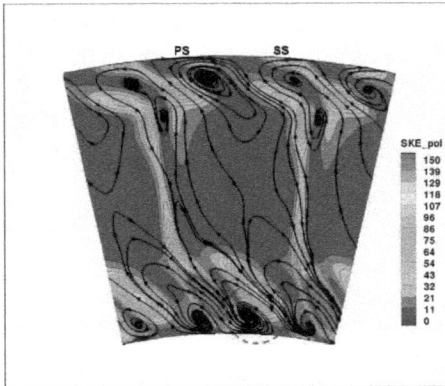

Figure 4.22: SKE distribution at stator exit plane with streamlines of the secondary velocity - Stator org. at design conditions from [76]

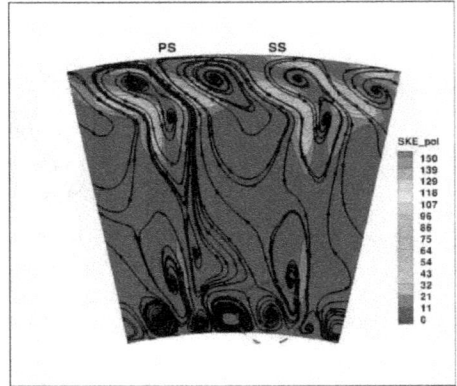

Figure 4.23: SKE distribution at stator exit plane with streamlines of the secondary velocity - Optimization A at design conditions from [76]

the pattern indicates a reduction in SKE not only in the hub region but also in the spanwise direction, showing a damping influence on the stator wake. Also, the vortex in the red dashed circle (which could display the passage vortex according to its sense of rotation) was moved further to the suction side than for the initial case. However, in the tip region no notable change in SKE distribution was observed.

4.3.3 Analysis of Optimization B - Shroud Design

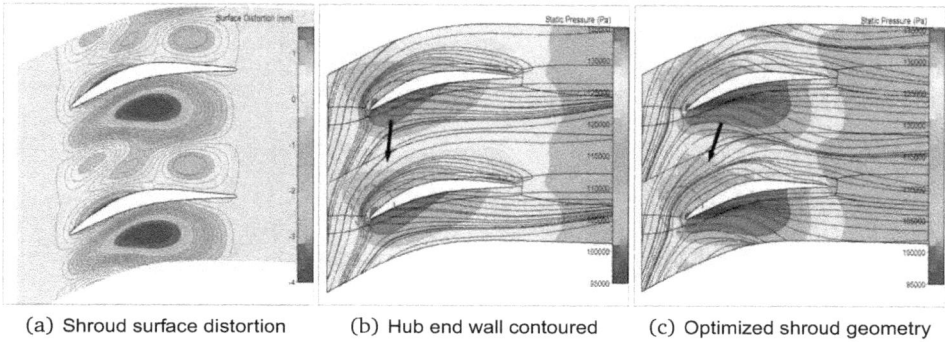

(a) Shroud surface distortion   (b) Hub end wall contoured   (c) Optimized shroud geometry

Figure 4.24: End wall height contour of optimized casing (a); Comparison of static pressure and streaklines on shroud surface (b, c) near stall from [76]

In Figure 4.24a the end wall shape of the casing after Optimization B is shown as a contour of surface distortion relative to the axisymmetric end wall. Like the hub end wall shape after Optimization A, the contour plot again shows a strong depression of about -4mm (5.7% span) next to the pressure side and a raised region in the rear part of the blade channel which is located much closer to the suction side compared to the profiled hub design. One smaller increase is also observed in the middle front part of the passage. Furthermore the optimization led to a rise of more than 1mm (1.4%span) at the early the suction side. As in the previous case, the contradictions of the lowering next to the pressure side with the associated stronger transverse static pressure gradient can be explained by examining the contours of static pressure together with the streaklines on the shroud surface (Figures 4.24b and 4.24c). In Figure 4.24b the recirculating particle paths indicate the suction side separation area near stall for the geometry from Optimization A. Comparing the static pressure contours, we can see that the static pressure on the pressure side was again remarkably increased by end wall contouring which corresponds to the depression in Figure 4.24a. The lower static pressure at the early suction surface is also in agreement with the decreased contour in this area. Again, the cause for all theses observations can be explained by the changed pressure distribution. The static pressure gradient is now higher in the circumferential sense but its effective direction has changed. Regarding the contoured shroud, the static pressure gradient is almost parallel to the streaklines in the throat area region and thus the effective force pushing the cross flow from pressure to suction side is reduced. At the leading edge the streaklines are further away from the suction side (marked with red dashed circle). In contrast, the static pressure gradient is more perpendicular to the streaklines in the rear part of the channel, leading again to overturning by encouraging the cross flow (Figure 4.18b). Overall, the changes in the flow field are in line with the ones observed for the hub optimization and therefore also the flow mechanism that prevents corner stall is the same. Referring back to Figure 4.14, the loss core (black dashed circle) has migrated even further from the end wall than it did previously for the hub optimization.

Figures 4.25 and 4.26 compare the SKE distribution of the design with hub contoured only and the design with both end walls optimized one axial chord length behind the stator exit. The mass

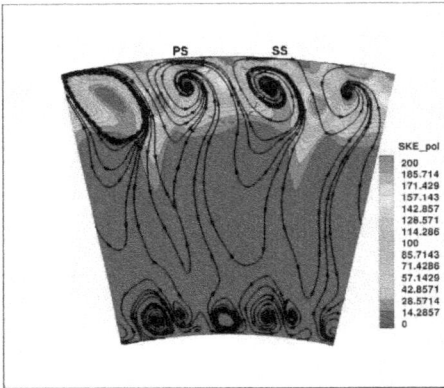

Figure 4.25: SKE distribution at stator exit plane with streamlines of the secondary velocity - Optimization A near stall from [76]

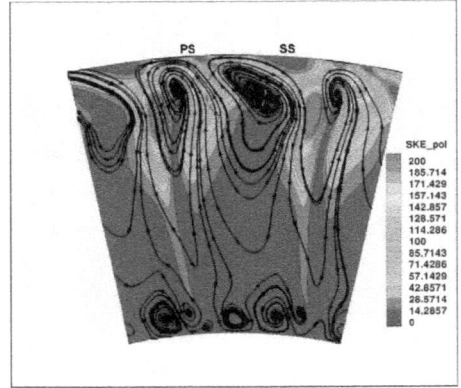

Figure 4.26: SKE distribution at stator exit plane with streamlines of the secondary velocity - Optimization B near stall from [76]

averaged SKE value did not change considerably, but the distribution did. The region of high SKE in the middle of the blade channel (red dashed circle) was compressed in the circumferential sense, but stretched radially. The core shows a higher concentration of SKE but over all, less area in the blade channel is covered. As in the SKE distribution of the design from Optimization A, this corresponds with the smaller separation bubble. The second region of high SKE, located left from the red circle, experienced the same changes. Both regions seem to squeeze the vortex that is located closer to the PS. The vortex was stretched in the radial direction. On the other hand, the vortex closer to the SS can now expand more in the circumferential direction and declined counterclockwise.

5 Rotor Optimization

The previous chapter focused on the separation phenomena in the stator row which is one of the disadvantages of the trend towards high stage loadings in the development of axial aircraft compressors. A mechanism was identified that suppresses corner stall by enhancing the cross flow in the rear part of the blade channel, resulting in a stronger overturning in the boundary layer. This is actually expected to lead to higher secondary flows. Since corner stall is the main source of loss in the stator, no final conclusions about the potential of end wall profiling to reduce secondary flow in a transonic compressor stator row could be made.

In contrast, the concern of this chapter lies on the application of the optimization chain to the rotor hub end wall of Configuration I of the Darmstadt Transonic Compressor. A general outline of this configuration was already given in chapter 4. In contrast to the original stator geometry, the original rotor design hardly shows separation regions at the hub end wall and blade suction side. The losses in the rotor are not dominated by separation. Hence, this study with a well-designed blade was expected to lead to a deeper understanding of how secondary flows in a transonic compressor can be influenced by the application of non-axisymmetric end walls. Table 5.1 shows the main aerodynamic design parameters for the rotor blade. Several optimization strategies involving different objective functions to be minimized and the corresponding performances were compared. Performance has to be seen in a global context and is not only limited to efficiency but has also to consider the total pressure rise. To specify this, it would be easy to reach a higher efficiency using an optimization if the blade profile was unloaded at the same time. If we just take a stator row, these two aspects do not conflict with each other since no work is introduced and an increase in efficiency always involves less pressure losses and therefore a higher total pressure ratio in the entire stage. In a rotor row, however, one has to consider that this conflict could lead to a lower stage pressure ratio at a higher efficiency, which would not be very surprising but is absolutely undesirable. The parameters considered within the optimization process were, hence, isentropic stage efficiency, pressure loss in the rotor, throat area and secondary kinetic energy (SKE). The rotor hub end wall of Configuration I of the Darmstadt Transonic Compressor was optimized within a single-point optimization at design conditions. For the datum design of this study a configuration coupled with the optimized stator was used. Moreover, a method to display secondary flows in turbomachinery was introduced. For this

Rotor design parameter	Hub	Mid	Tip
Turning [°]	24.27	10.23	3.32
ΔW_φ [m/s]	79.05	101.85	118.36
Chord length [m]	0.09	0.088	0.072
Stator solidity (chord/pitch)	1.87	1.48	1.21
Diffusion factor	0.36	0.37	0.38
Aspect ratio	1	0.98	0.8

Table 5.1: Aerodynamic rotor blade design parameters from Schulze [84]

method, a second CFD simulation is used to calculate the primary flow where the hub end wall is defined as an Euler wall to avoid the end wall boundary layer and so to eliminate the cause for some of the secondary flow mechanisms. This method clearly shows how the characteristics of secondary flow can be positively influenced by using non-axisymmetric end walls.

5.1 Review on Rotor Design Study with Original Stator Design

This section addresses the main differences compared to the rotor design study which was presented at the ASME TurboExpo in 2009 [77]. Table 5.2 provides a summary of this first rotor design study. The parameter variation within the optimizations resulted in the following observations: Strong penalties on SKE at the rotor outlet and moderate penalties on isentropic efficiency, throat area and pressure ratio led to the best design. Isentropic efficiency could be raised by 0.12%, SKE at the rotor exit was reduced while the total pressure ratio of the stage remained constant. Strong penalties on efficiency and pressure ratio, a moderate one on throat area and a small one on SKE at the rotor outlet all led to a smaller increase in efficiency: 0.06%.

	η	PR	η_{Rotor}	CP_{Stator}
Original geometry	85.37	1.4665	89.92	0.0895
Rotor design focused on η	85.43	1.4691	89.95	0.0889
Rotor design focused on SKE	85.49	1.4662	89.98	0.0876

Table 5.2: Summary of rotor optimization study with original stator design

The reason for the contradiction, that a focus on SKE reduction finally lead to a higher gain in efficiency than the optimization which had actually directed at an increase in efficiency can be explained by the aforementioned conflict to the total pressure ratio. Table 5.3 and Table 5.4 show the optimization settings of this first rotor design study. The total pressure ratio of the datum design (with the poor stator design) was found to be $PR_{Total} = 1.4665$. In order to investigate the optimizer's capability to find an alternative design with both increased efficiency and total pressure ratio, the required value for the latter was set to 1.5. However, by applying different penalties on the pressure constraint, i.e. 34% influence for the optimization focused on gain in efficiency and 10% for the optimization focused on reduction in SKE, the first optimization ended exactly in this conflict by showing an increased total pressure ratio. On the other hand, the SKE-optimization resulted in a higher rise in efficiency while the total pressure ratio almost remained at the original value. This clarifies again, that choosing the proper objective function and the optimization targets is a highly sensitive issue.

The second matter of this first design study concerns to what extent the rotor and the stator rows are responsible for the improvements. According to the stator pressure loss coefficients in Table 5.2, improved stator inflow conditions are responsible for a quite large portion of the better performance. Since the original stator design shows a huge corner separation, small changes in the rotor-outflow/stator-inflow conditions caused by profiled end walls may result in a positive modification of the separation. Therefore, the optimized rotor configurations alone are not responsible for the improvements, which is also confirmed by the values of the rotor-only efficiency in Table 5.2.

	η [%]	PR_{Total} [-]	SKE [m^2/s^2]	Throat area [m^2]
Required value	100	1.5	0	0.002888
Weighting factor	900	3000	$2.5 * 10^{-5}$	30000
Weighting in percent $w_i P_i / OF$ [%]	51	34	5	10

Table 5.3: Objective function ASME study - focus on efficiency [77]

	η [%]	PR_{Total} [-]	SKE [m^2/s^2]	Throat area [m^2]
Required value	100	1.5	0	0.002888
Weighting factor	200	2000	$1.5 * 10^{-3}$	30000
Weighting in percent $w_i P_i / OF$ [%]	12	10	69	9

Table 5.4: Objective function ASME study - focus on SKE [77]

5.1.1 Numerical Setup

For the rotor optimization, the numerical procedure was identical to the one that was used for profiling the stator end walls. All parameters taken from the described parameterization method were allowed to vary between -5mm and 5mm in the vertical direction. With an average rotor span of 85mm, this corresponds to approximately 6% span. For a description of the grid and the computational methodology, please refer to chapter 4. Although the objective was to optimize the rotor, the stator row was included in the full-stage CFD simulations.

5.2 Settings of the End Wall Optimizations

The end wall shape of the rotor hub was designed through a single-point optimization at design speed of 20,000rpm. The optimization was carried out at a reduced mass flow rate of 16.1kg/s where the compressor (with the previously optimized stator) has a peak-efficiency of 87.22%. The optimization was intended to be carried out at constant mass flow rate to remain at the same operating point throughout the whole procedure. A change in mass flow (e.g. by keeping the static pressure pressure at the exit constant) would lead to difficulties in the comparison of the results of the initial and optimized geometries. Indeed, for a correct comparison of two designs, the mass flow and the total pressure ratio have to be constant. If an optimization leads to a higher total pressure ratio, one has to calculate the isentropic efficiency that the new design would have at the datum pressure ratio by means of the polytropic efficiency. The objective function of the design process contained four terms: isentropic stage efficiency, absolute total pressure ratio, SKE at the rotor exit plane just in front of the trailing edge and the throat area [76].

$$OF = w_\eta P_\eta + w_{PR\_abs} P_{PR\_abs} + w_{SKE\_out} P_{SKE\_out} + w_A P_A \qquad (5.1)$$

In order to take advantage of the experience from the first rotor design study, the required value for the pressure ratio was set to the original value of 1.48. Thus, the corresponding penalty term is only valid if the pressure ratio of a sample is below this value. Otherwise the penalty term equals zero (side condition), reducing the objective function to three terms. In general, it is the easier for an optimization algorithm to find a better design the fewer constraints are applied. Although a broader variation of objective functions was investigated, only the ones that led to the most interesting results will be discussed. Table 5.5 summarizes the required values, weighting factors and the resulting influence in percent of the penalties on the optimization that was focused on isentropic efficiency. Table 5.6 does the same for the optimization focused on SKE at the rotor exit.

	η [%]	PR_{Total} [-]	SKE [m^2/s^2]	Throat area [m^2]
Required value	100	1.5	0	0.002888
Weighting factor	900	3000	$2.5 * 10^{-5}$	30000
Weighting in percent $w_i P_i / OF$ [%]	51	34	5	10

Table 5.5: Parameters for the objective function - main focus on efficiency

	η [%]	PR_{Total} [-]	SKE [m^2/s^2]	Throat area [m^2]
Required value	100	1.48	0	0.002888
Weighting factor	600	40000	$1.55 * 10^{-3}$	40000
Weighting in percent $w_i P_i / OF$ [%]	15	14	65	6

Table 5.6: Parameters for the objective function - main focus SKE at rotor exit

5.3 Optimization Results - Overview

This section gives an overview of the results and the observed phenomena. Explanations of the improvements in efficiency and overall performance are given in the following sections. Figure 5.1 compares the isentropic efficiency versus normalized mass flow for the datum axisymmetric design and the two optimized designs (points within the dashed circle) with contoured hub end walls. The benefits observed are rather small and even smaller than for the first rotor design study [77]. The latter was expected due to the absence of large separations in the stator row, for which reason, modified stator inflow conditions would not significantly contribute to the improvements. One can also notice that the more elaborate choice of the objective function resulted in the intended optimization targets. The optimization that focused on efficiency led to the higher gain in efficiency in the range of 0.07% for the design point whereas the optimization that focused on reduction of SKE at the rotor exit plane only shows a very tiny benefit of around 0.03%. Although not shown here because of the difficulty in visualizing the small differences, it should be noted that all contoured geometries reached the original numerical stability limit. Figure 5.2 shows the absolute total pressure ratio for all designs. In contrast to a non-rotating row or a cascade without work input where an increase in efficiency also involves a raise in pressure ratio, both parameters need to be considered to evaluate a design.

It is simple to gain a higher efficiency when the profile is unloaded at the same time and the rotor builds up a smaller pressure rise. Then, of course, the losses will also decrease, leading to a higher efficiency. This should be avoided by applying a penalty on the total pressure ratio.

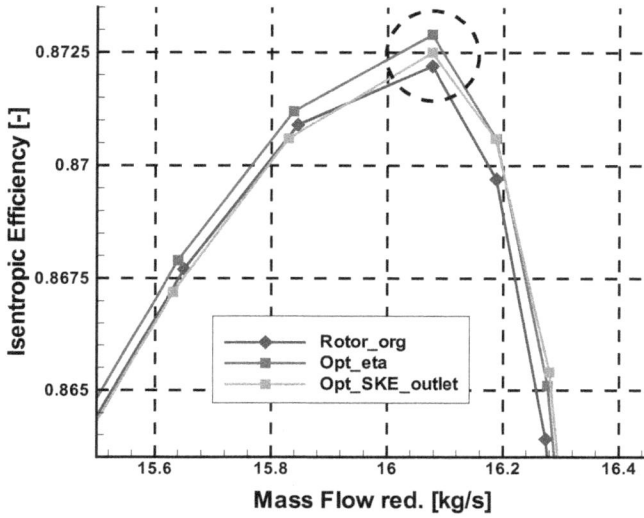

Figure 5.1: Characteristic lines of datum and optimized designs - isentropic efficiency

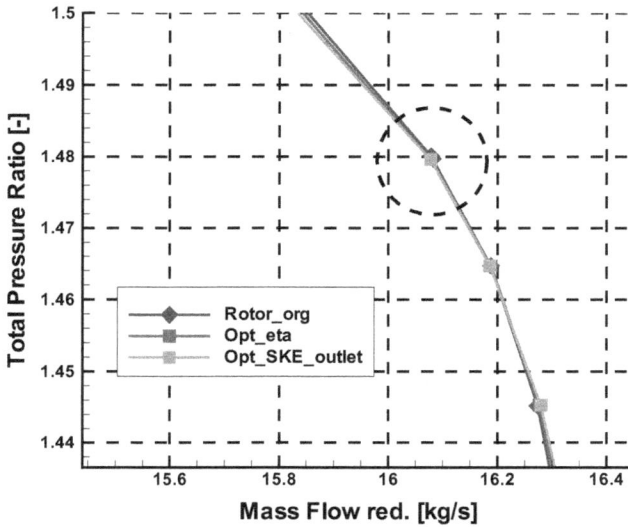

Figure 5.2: Characteristic lines of datum and optimized designs - absolute total pressure ratio

Computation	η [%]	PR_{Total} [-]	Throat area [m^2]
Datum design	87.22	1.480	0.002888
Optimization - focus efficiency	87.29	1.479	0.002890
Optimization - focus SKE outlet	87.25	1.479	0.002891

Table 5.7: Performance comparison between original geometry and optimized designs

According to Figure 5.2, this was successfully accomplished and the values of the pressure ratio match the datum level for both optimizations.

The values of the most important parameters are summarized in Table 5.7 for all designs. The changes to efficiency are sufficiently small that some comments shall be given why the author is confident that the deltas do not fall within the 'error bars' of the computation. Besides the changes of the flow field which will be discussed in the next sections, both the simulations of the original design and the optimized designs were run on a second grid to estimate the accuracy. This mesh is slightly finer than the grid used for the optimization. It consists of 2.4510^6 grid points (compared to 2.1510^6 of the datum grid) with 1.510^6 points for the rotor row (compared to 1.3510^6 points) and 935,000 grid points for the stator row (compared to 800,000 points). Exemplarily, the results of the datum design and the efficiency-optimized design are presented. The computations with the second grid led to an isentropic efficiency of 86.88% at a TPR of 1.495 for the axisymmetric design compared to 86.94% at a TPR of 1.494 for the optimized design. These values differ from the results obtained with the datum mesh. However, the delta in efficiency is with 0.6% in the same range as for the optimization with the coarser grid. Table 5.7 shows that the throat area of the compressor was only slightly increased during the optimization and should thus have a negligible impact.

Figure 5.3 shows the radial distribution of the relative pressure loss coefficient for the rotor, given by

$$CP_{rel} = \frac{p'_{rel\_out}(r) - p_{rel\_out}(r)}{p_{dyn\_in}(r)}. \tag{5.2}$$

Due to the rising hub in axial direction, the matching radial positions at inlet and outlet have to be evaluated according to the corresponding relative channel height. p' is the rothalpic isentropic pressure at rotor outlet that would appear in the ideal case without any losses. This correction by the rothalpy is necessary since the radii of the streamlines increase between inlet and outlet which alone leads to a rise in total relative pressure due to the change of the local blade speed. This is desirable because the part of the enthalpy rise attributable to the change in blade speed is essentially loss free [9]. For a detailed derivation of the considered pressure loss coefficient, please refer to the appendix.

According to Figure 5.3, it is remarkable that the non-axisymmetric end wall contours have a negative influence on the region adjacent to the end wall, up to 5% channel height, although the initial intention was to improve just this area. This is slightly more pronounced for the optimization focused on efficiency, which led to the largest overall benefits. On the other hand, the core stream is positively influenced by the contoured end walls. Here, both optimized designs show approximately the same reduction in pressure loss, although the decrease is a little higher for the efficiency-optimized design between 10-15% channel height.

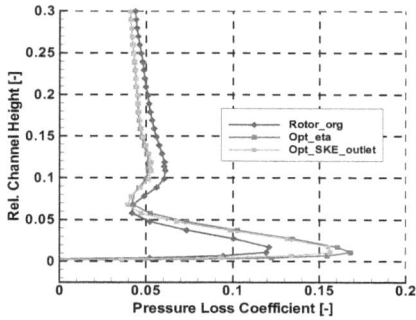

Figure 5.3: Radial distribution of the relative pressure loss coefficient

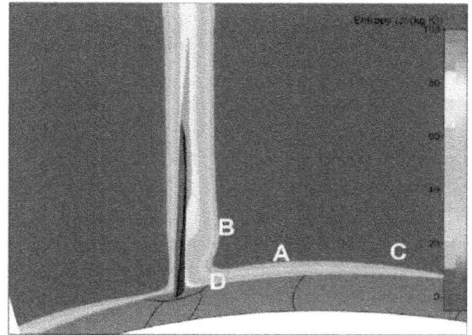

Figure 5.6: Entropy distribution at rotor exit - datum design

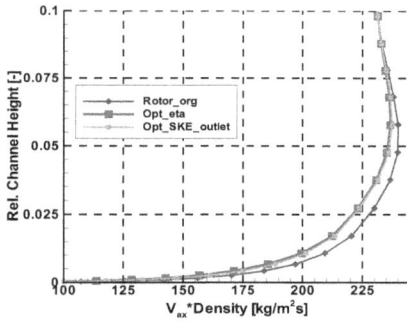

Figure 5.4: Radial mass flow distribution in the end wall region up to 10% span

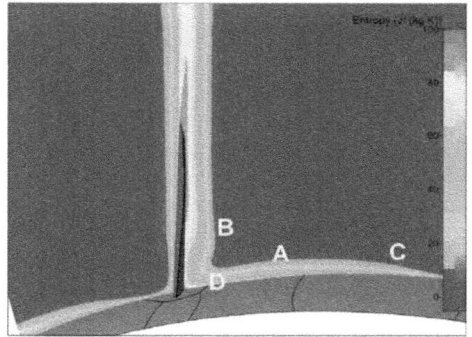

Figure 5.7: Entropy distribution at rotor exit - SKE-optimized design

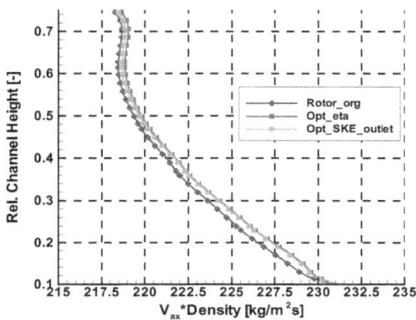

Figure 5.5: Radial mass flow distribution in the free stream

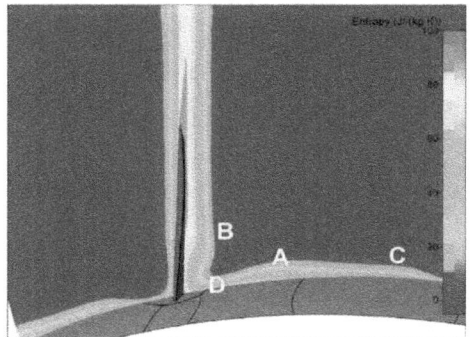

Figure 5.8: Entropy distribution at rotor exit - efficiency-optimized design

Since this is a transonic rotor, changes in end wall blockage can of course affect the 2D flow. This is confirmed by the data presented in in Figure 5.4, which shows the radial mass flow distribution in the end wall region up to 10% span. The end wall profiling increases end wall blockage which leads to smaller flow rates for all contoured designs in this region. Therefore, more mass flow is shifted into the outer regions as can be seen in Figure 5.5. Compared to the SKE-optimized design this general tendency is more pronounced for the design focused on efficiency, especially in the region between 45% and 75% span. Apparently, the best end wall design is the one that produces the most blockage at the expense of near hub end wall loss. This is then overcompensated as this design also shifts most of the mass flow to the low-loss free stream which results in an overall gain in efficiency. This observed phenomena is similar to the effect resulting from the application of compound lean. The introduction of lean also intends to shift more mass flow towards the mid sections by additional radial forces. However, the big difference is, that leaned blades are usually characterized by improved end wall sections and slightly higher losses at mid-span [36], which is not the case here.

In order to gain a deeper understanding where these differences in loss occur, Figures 5.6 to 5.8 show the entropy distribution at the rotor exit plane for the original and the optimized designs. The figures confirm the evaluation of the radial profiles but additionally show where the losses are produced in the channel. The views are oriented upstream, symbol B marking the suction side of the blade. As expected, the local changes in entropy are very small and hard to interpret. The datum design shows higher losses at the suction side. The yellow region next to the symbol B could particularly be reduced by both optimizations. On the other hand, the optimized designs produce more losses in the end wall region close to the suction side (symbol A). This latter effect seems to be more pronounced for the SKE-optimized design, especially as there is a second smaller region of higher entropy towards the pressure side (symbol C). An explanation why the first optimized geometry, nonetheless, shows higher circumferentially averaged total pressure losses in the end wall region may be given by the area of lower losses close to the corner of the trailing edge (symbol D) for the second optimization. The loss profile in the remaining area across the channel looks very similar.

Figures 5.9 and 5.10 show the evolution of mass-averaged SKE along the axial extension of the blade channel, based on radial velocity profiles and Euler walls. The plane at normalized blade channel=1 was used for the optimization. Both sets of curves show varying characteristics. In Figure 5.9, the SKE decreases in the first 20% of the channel which is quite unexpected since the flow turning actually increases in the first 30% of the blade channel and the secondary flow would therefore also be assumed to increase. After the peak at around 40% channel length, both optimized designs show a higher SKE than the datum design. Only one axial position upstream from the trailing edge, the optimization, focusing on efficiency, shows a reduction. At the trailing edge, the values of both optimized designs are smaller than for the original axisymmetric design.

SKE increases up to 30% axial chord length using Method 2 (Figure 5.10). After the peak, only the SKE-optimized design features a significantly smaller level in the rear part of the blade channel. The other optimized design initially even shows higher SKE-values around the peak and stays more or less at the level of the original rotor after that. At the last rotor plane, all optimized designs show a reduced SKE value. Overall, Figure 5.10 shows a more realistic distribution than Figure 5.9. The main reason therefore is based on the fact that the potential field of the blade has not been filtered out in SKE Method 1. According to this deficit, the potential flow field also contributes to the secondary flows and therefore to SKE. This is not the case in the theory on secondary flows. Fortunately, a reduction of SKE could be observed for the

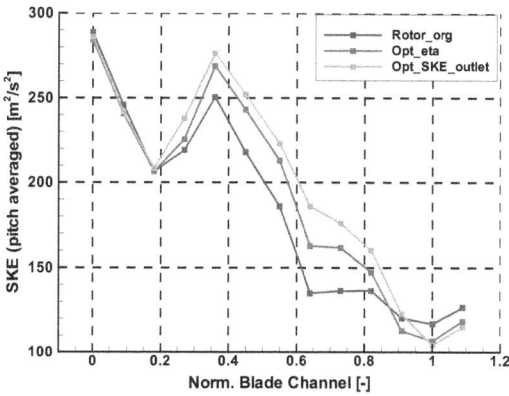

Figure 5.9: Mass-averaged SKE (based on radial velocity profiles) along normalized blade channel according to Method 1

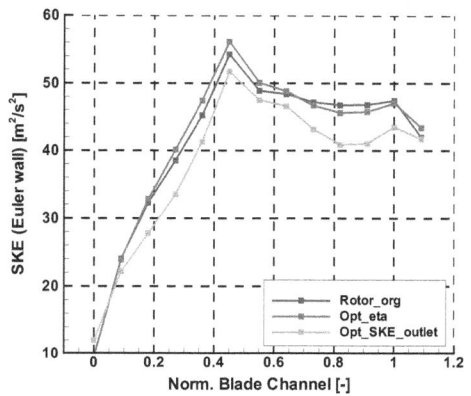

Figure 5.10: Mass-averaged SKE (based on Euler walls) along normalized blade channel according to Method 2

corresponding optimization in both sets of curves in the rear part of the blade channel, i.e. the part that has been chosen for the evaluation of the designs during the optimization. A focus on sections in the first 80% of the blade channel would probably have led to inadequate designs when using the first Method. For that purpose, the definition based on Euler walls should be used in the optimization process, along with awareness that this method is much more expensive in terms of CPU-time due to the additional CFD computation required.

5.3.1 Analysis of Improvements

First of all, it will be checked, whether the application of contoured end walls has an impact on the shock system of the investigated compressor configuration and compared to literature. Figure 5.11 displays the isentropic Mach number at 5% channel height. One considerable difference is that the suction side peak Mach number of the datum rotor appears smoothed out for both optimized designs. Overall the deceleration of the fluid on the suction side is quite constant and according to this characteristics there is no strong shock system visible at this blade height. To confirm this observation, the relative Mach number distribution in the blade channel at the same span is plotted in Figures 5.12 and 5.13 for the original rotor and the SKE-optimized design, respectively. The latter plot is representative for both optimizations as they hardly differ. The black line represents the isoline of $Ma = 1$ and proves the supersonic region to be locally very restricted. Even more important is the fact, that this area does not cover the entire blade channel avoiding the cross section to be chocked with the resulting strong shock pattern.

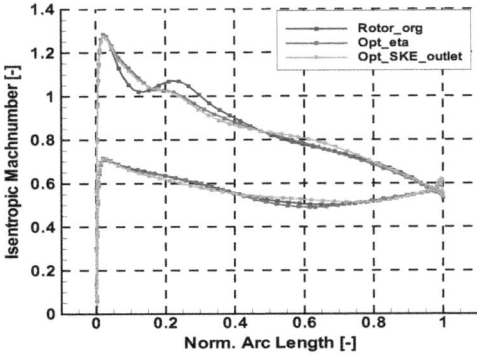

Figure 5.11: Isentropic Mach number distribution at 5% channel height

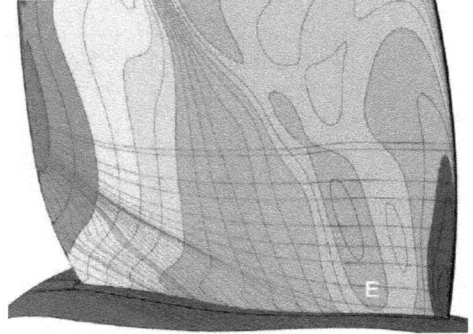

Figure 5.14: Streaklines and static pressure on the rotor suction side - datum design

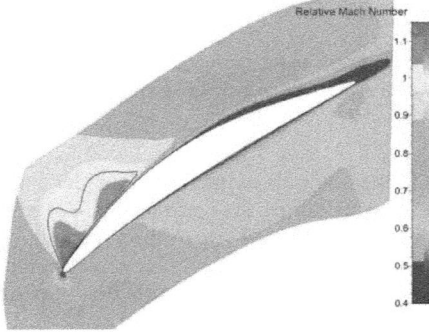

Figure 5.12: Relative Mach number distribution at 5% channel height - datum design

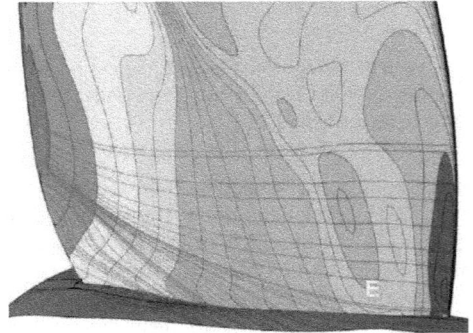

Figure 5.15: Streaklines and static pressure on the rotor suction side - optimized design (efficiency)

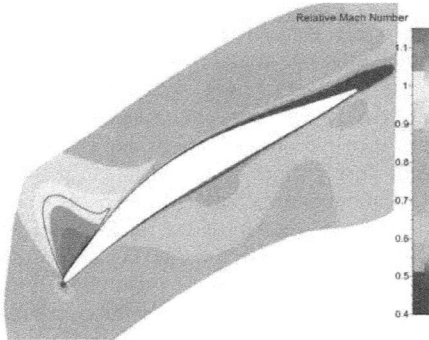

Figure 5.13: Relative Mach number distribution at 5% channel height - SKE-optimized design

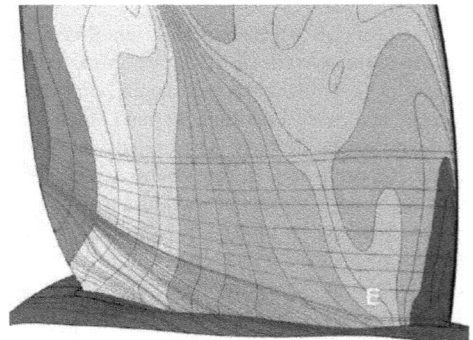

Figure 5.16: Streaklines and static pressure on the rotor suction side - optimized design (SKE)

5 Rotor Optimization

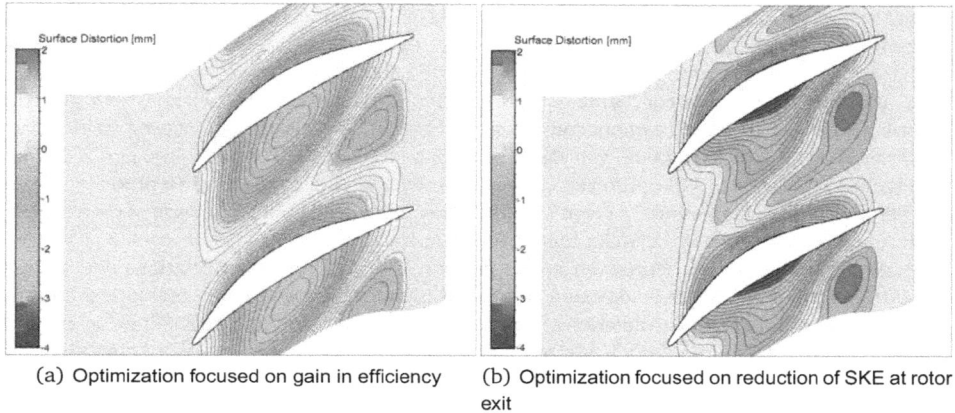

(a) Optimization focused on gain in efficiency

(b) Optimization focused on reduction of SKE at rotor exit

Figure 5.17: End wall height contours of optimized hub designs

This was to be found quite different in the work of Iliopoulou et al. [52] who obtained an extremely modified shock pattern near the hub. Their shock pattern had covered the entire blade channel before the hub was contoured. Referring back to Figure 5.11, there is no noticeable difference in terms of global loading (pressure difference between pressure and suction side). However, both optimized designs demonstrate a slightly higher blade load from 0.05-0.15 blade chord length, whereas the datum design shows a higher loading downstream until 0.45 chord length. Since the onset of secondary flow does not only depend on the global blade load but also on its distribution, the datum design should produce stronger secondary flows due to the larger extension of this area.

This assumption is enforced by Figures 5.14 to 5.16 which show the static pressure distribution on the rotor suction side together with oil traces. All designs have in common, that the corner vortex starts to roll up and lift off along the suction side more or less at the same axial position. However, if we follow the main trajectory (merging streaklines) of the vortex, we can notice, that this vortex leaves the blade passage at a lower radial position for the optimized designs and the area of higher static pressure (indicated by green contours) starts closer to the leading edge. This is particular true for the SKE-optimized design. Although the chordwise expansion of the low pressure region is comparable to the datum design for the sample based on efficiency gain, the tongue of low pressure (symbol E) is much smaller in that case and does not reach the hub. This reduction alone seems to be sufficient to smooth the blade loading characteristics.

In Figures 5.17a and 5.17b, the end wall shapes of the optimized designs are shown as contours of surface distortion relative to the datum axisymmetric end wall. There are mainly three common changes to be observe for both optimized geometries which are more or less distinct. Both contour plots feature a strong depression located next to the pressure side. This depression extends tongue-shaped into the middle of the blade channel. A smaller dip can be observed at the suction side. Both Figures show additionally a rise in the rear part of the blade channel. This rise has its peak almost at the trailing edge and seems to separate the two depressions from each other further upstream in the channel. In Figures 5.7 and 5.8, these raised areas coincide with the higher loss production in the boundary layer at symbol A and seems therefore to be responsible for the higher blockage in the end wall region. This analysis is supported as Figure 5.17b alone features a strong peak within the elevated area. The circumferential position of this peak corresponds to the second area of increased blockage in the boundary layer in Figure 5.8 (symbol C). Nevertheless, it can not completely be clarified why the design focused on efficiency shifts more mass flow to regions even above mid-span (see sections between 45% and 75% span in Figure 5.5) than the SKE-optimized design, even though the latter seems to produce slightly higher blockage in the boundary layer. One the other side, one should be aware of the very tiny differences in performance between the two geometries.

However, the observed depression at the pressure side is against expectations and turbine research experience as this is assumed to lead to higher transverse static pressure gradients. To help understand the contradictions, contours of static pressure and streaklines on the hub surface are displayed in Figures 5.18a to 5.18c. Via macro, the streaklines were guaranteed to begin at the same starting points for all particle paths. In comparison to the datum design, we can see that the static pressure on the pressure side was increased, in line with the assumption, only for Figure 5.18c. In Figure 5.18b, the size of the region with high static pressure remained constant. But, the area of higher static pressure was moved slightly upstream towards the leading edge. Nonetheless, improvements could be achieved for these designs because of the changed static pressure distribution.

The black arrows indicate the effective direction of the force induced by the static pressure. This direction considerably changed, depending on the optimization case. The pressure isolines are much more perpendicular to the streaklines on the suction side at the leading edge for both optimized designs, reducing thus the effective force that pushes the flow towards the suction surface.

In both Figures 5.18b and 5.18c, one can observe that the streaklines impact on the suction side further downstream. Especially for Figures 5.18c this happens with a much smoother angle. The small lowering at the suction side is believed to be responsible for that change. Both optimized designs also show a remarkable change of the static pressure force in the rear part of the suction side. This alternating pressure pattern is a bit less pronounced in Figure 5.18b. Comparing this observation to the obtained contours, the rise in the rear part is the cause for this change as the contour related to Figure 5.18b additionally shows a bump within the raised region in the rear part. Finally, the altered pressure distribution in the middle of the blade channel (indicated by the arrow closest to the pressure side) is obviously based on the depression at the pressure side which reaches into the mid-channel. This is more distinct for the SKE-optimized design and also explains the improvements concerning Figure 5.18c, although the circumferential static pressure gradient has increased. The depression prevents an increase in the static pressure perpendicular to the flow direction in this area. According to the streaklines, the rise in the rear part leads to a smoother deflection of the flow whereas the other two features together result in an streakline impact that is located further downstream at the blade suction side. The described observations suggest that a strong development of the new features leads to a higher reduction of secondary flow and secondary losses whereas the largest improvements in terms of efficiency can be obtained with moderate characteristics concerning the new end wall design.

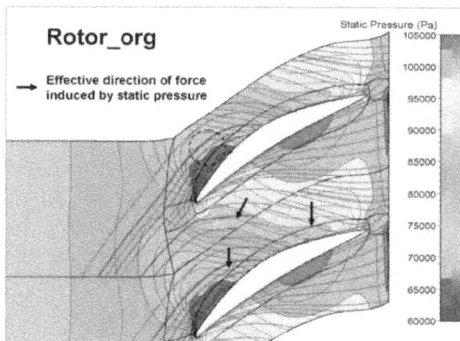

(a) Datum design

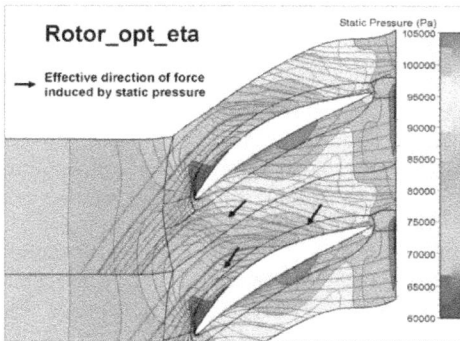

(a) Datum design

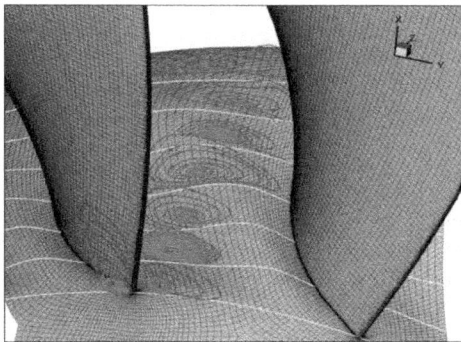

(b) Optimization focused on gain in efficiency

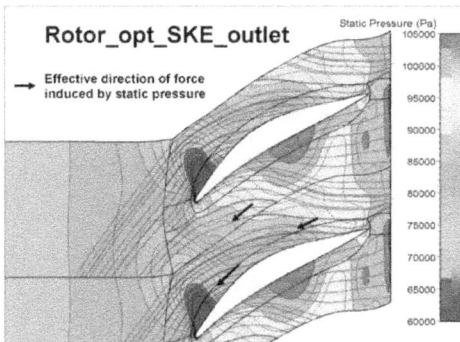

(b) Optimization focused on gain in efficiency

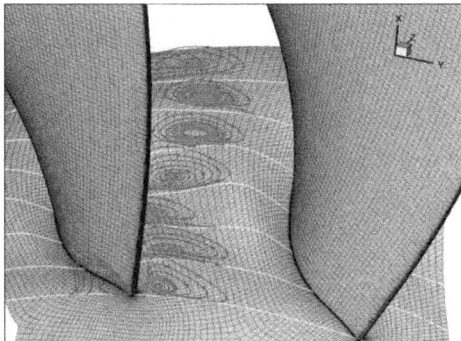

(c) Optimization focused on reduction of SKE at rotor exit

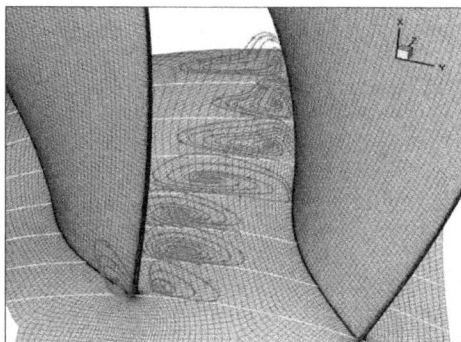

(c) Optimization focused on reduction of SKE at rotor exit

Figure 5.18: Comparison of static pressure distribution and streaklines on hub surface

Figure 5.19: Visualization of horseshoe vortices and passage vortex formation based on Euler walls

5.3.2 Visualization of secondary flows

In Figures 5.19a to 5.19c, the secondary flow features based on the Euler-wall-method are visualized using secondary streamlines at defined constant axial positions. The observations in this section are limited to this method since the visualization based on radial velocity profiles was quite unsatisfactory. In all graphics, the suction side and the pressure side legs of the uprolling horseshoe vortex rotating in opposite directions are clearly visible. It also becomes obvious how the pressure side leg emerges as the core of the passage vortex being driven towards the suction side in the downstream direction. In the original design in Figure 5.19a, the passage vortex is driven across the entire blade channel and impacts on the suction side of the trailing edge. The radial position of the impact corresponds to the position of symbol B in Figure 5.6 and explains the higher losses in the boundary layer of the suction side at the rotor exit. The deflection of the passage vortex was reduced for both optimized designs, whereas the cross flow could be counteracted most in Figure 5.19c. It does not impact the suction side and thus produces less losses on the suction side. Comparing both new designs, the optimization focused on SKE reduction (Figure 5.19c) shows the weakest deflection of the passage vortex. This is in line with the progress of the streaklines and confirms the presented assumption from the previous section that geometry changes should be large to influence the development of the secondary flow. For the design in Figure 5.19b (although showing a weaker deflection of the passage vortex compared to the datum design), the outer secondary streamlines of the vortex core impact onto the suction side and therefore produce slightly higher losses at the trailing edge in comparison with the SKE-optimized design (see symbol D in Figure 5.7 and 5.8). If the results shown in Figure 5.10 are considered again, the reduction in the integral SKE in the rear part of the blade channel does not seem to be the main effect as it could only be achieved for one of the optimizations. It is more important to note that both optimizations led to a reallocation of the SKE, visible by the evolution of the passage vortex core along the channel.

6 Unsteady Investigations

This chapter describes the unsteady analysis of the original and optimized stator designs. An investigation of the optimized rotor designs in unsteady mode has not been carried out due to the small differences and improvements in performance that have been observed during the rotor optimization study. Since every unsteady simulation, once converged, is characterized by a periodically oscillating solution it had been expected to be very challenging if not entirely impossible to analyze differences in efficiency of $\Delta\eta \approx 0.05\%$, the amplitude of the oscillation being in the range of 0.4% at the same time.

As already mentioned in section 3.3.4, the domain scaling method with a sliding mesh approach was employed to conduct the unsteady simulation. As quite common in turbomachinery applications, the blade matching of rotor (16) and stator rows (29) required a scaling of the geometry as a simulation of higher numbers of blade passages was not suitable for this configuration. In order to obtain the constraint of identical pitch distances the number of stator blades was increased to 32 which enabled to set up unsteady simulations with one rotor passage and two stator passages. To minimize the error, which would be unavoidably introduced by increasing the blade number, the pitch-to-chord ratio of the stator row was modified by the factor 29/32 respectively. The shorter chord length was expected to compensate the smaller aerodynamic blade load which would be introduced due to the higher number of blades.

After the geometry scaling, the aerodynamic similarity of both original stator configurations (scaled and unscaled) was to be validated in terms of overall performance (speed lines), blade loading (isentropic Mach number) and deceleration of the fluid (diffusion factor). The study was part of a student thesis [57] in which the interested reader will find detailed results of the comparison. At this point, only a rough summary of the conclusion will be given. In the study, the scaled optimized stator (32 blades) turned out to exactly show the same behavior in all three criteria as the unscaled optimized stator (29 blades). However, the scaled original stator design showed a slightly better behavior than the unscaled stator. Especially, the isentropic Mach number distribution indicated a smaller separation area for the scaled design. Therefore, the aerodynamic relief by increasing the blade number seems to overcompensate the additional load by reducing the chord length if separation is present.

Nonetheless, it should be made clear that these differences were found to be very small and in a tolerable range. Hence, there was confidence to use the scaled design for both the datum and optimized stator configurations to carry out the unsteady investigations and to draw conclusions for the unscaled designs. All comparisons to steady solutions in this chapter will therefore be related to the scaled stator geometry with 32 blades.

6.1 Choice of the Time Step

To estimate the suitable time step which would be needed to capture all relevant phenomena for the investigated compressor configurations, the criterion of Gourdain and Leboeuf [28] (see section 3.3.3 for details) was adapted to the used mesh density and the rig parameters.

In a first step, the smallest circumferential dimension of the mesh l_ϕ was evaluated at the regions of interest which are near hub and near shroud. As an axial reference, the rotor-stator-interface was taken as the main interaction was expected to occur there. At the hub the mesh featured the smallest dimension of two adjacent cells with $l_\phi = 2.1mm$ and at the shroud with $l_\phi = 4.3mm$. According to equation 3.40 with correlation 3.38 and 16 rotor blades this results in values of $a_{max} = 22.44$ for the hub region ($r_{hub} = 120mm$) and $a_{max} = 17.35$ for the shroud region ($r_{shroud} = 190mm$). The blade passing frequency of the present rotor is

$$f_{BPF} = \frac{N_R \cdot rpm}{60} = \frac{16 \cdot 20220}{60} = 5392Hz \tag{6.1}$$

which enables the mesh to resolve frequencies up to $133,114Hz$ at the hub and, respectively, $93,551Hz$ at the casing. This means that the physical time to capture all frequencies is determined by the hub region and should be set to $\Delta t = 3.756 \cdot 10^{-06}s$ in this case.

In context of unsteady turbomachinery simulations, it is common to adjust the number of physical time steps per passage rather than to choose its dimension. A blade passing frequency of $f_{BPF} = 5392Hz$ corresponds to a physical time of $T_{pass} = 1.8546 \cdot 10^{-04}s$ for one rotor blade to pass. This leads to a required number of physical time steps of

$$k = \frac{T_{pass}}{\Delta t} \approx 50. \tag{6.2}$$

As some of the cells around the blade were of slightly smaller circumferential dimension the number of time steps for the simulations was increased to 60 per passage which corresponds to a physical time step of $\Delta t = 3.0910 \cdot 10^{-06}s$ in order to ensure to also capture these effects, particularly in terms of corner separation.

When performing an unsteady CFD calculation it is appropriate to start with very few inner iterations (so called pseudo time steps) for each time step in order to obtain a periodical signal within a short time. Once, that signal has been established the number of inner iterations should be increased to guarantee convergence for each physical time step. In this case, 10 pseudo time steps were used at the beginning which were later on changed to 50.

6.1.1 Convergence

Figure 6.1 shows the convergence history of the isentropic efficiency for the optimized stator design near stall. Also plotted as red crosses are the time-averaged values of the isentropic efficiency for each passage. This example will be representatively analyzed for all simulations. In this plot, one can observe a recurring pattern which repeats after 420 time steps which corresponds to 7 rotor passages having passed the stator row. For a really completely converged unsteady calculation one would expect to obtain the same sinusoidal signal with the same amplitude for each passage. However, and this is a quite important fact to justify the evaluation of such a simulation, the time averaged values of the efficiency remain at a constant level and do not alter against the rotor passages. Moreover, the amplitudes of the different rotor passages seem to decrease in such a way that almost a pure sinusoidal signal could have been expected if the simulation would have ran longer. Nonetheless, we will try to look into the reasons for the deviations from a ideal convergence history.

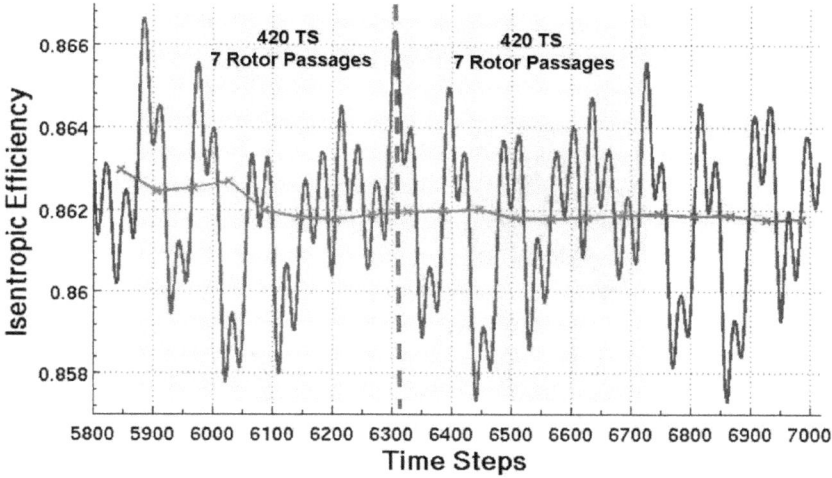

Figure 6.1: Convergence history for optimized stator design near stall

One of the main causes for such deviations are known to be disturbances which travel within the flow domain. If non-reflecting boundaries for inlet and outlet of the flow domain are not implemented in the employed CFD solver, these disturbances are likely to be reflected at the outlet and travel upstream towards the inlet and vice versa which leads to an overlapping long-wave superposition to the short-wave signal of the rotor passing frequency. Similar to the pressure information, the described disturbances are assumed to propagate with the speed of sound. If we take an average static temperature, e.g. at the rotor-stator interface, of $T = 315K$ this results in an average speed of sound in the flow domain of

$$a = \sqrt{\gamma R T} = 355 \frac{m}{s}. \qquad (6.3)$$

With a physical time of $420 \Delta t = 1.298 \cdot 10^{-03} s$ of the entire process we need to estimate the distance one fluid particle covers within the flow domain to confirm the assumption whether the observed disturbances arise from the domain boundaries or not.

Figures 6.2 and 6.3 display a representative stream surface at 50% channel height in the blade-to-blade and the meridional view respectively. As we can see in Figure 6.2, the fluid particle once it has entered the stator row does hardly experience any movement in the circumferential direction. It this therefore fair to neglect this portion and take just the axial distance of $l_{Stator} = 0.153m$ as a running length in the stator row. In contrast, the fluid particle passes a considerable distance in the pitchwise direction in the rotor row. The fluid particle needs approximately 5 rotor passages until it has reached the rotor-stator-interface from the inlet. If we take the average of the two representative radii at the inlet and the rotor-stator-interface ($r = 0.14m$) this leads to a pitch distance of

$$d = 2 \pi r \frac{5}{16} = 0.274m. \qquad (6.4)$$

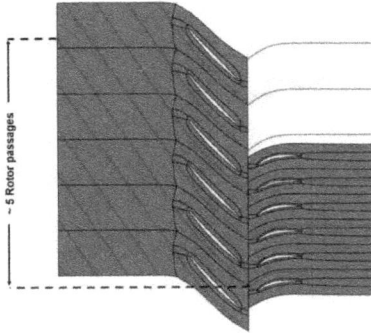

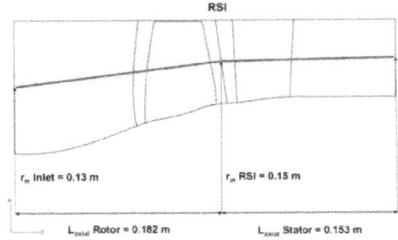

Figure 6.2: Estimation of distance a fluid particle travels from inlet to outlet in unsteady mode

Figure 6.3: Meridional view of stream surface at 50% channel height

The radial off-set of the representative stream surface of $\Delta r = 0.02m$ delivers together with the axial length of the rotor $l_{Rotor} = 0.182m$ a meridional distance of the stream line of

$$l_{m\_rotor} = \sqrt{\Delta r^2 + l_{axial}^2} = 0.183m. \qquad (6.5)$$

This leads to an entire distance in the rotor row of

$$l_{m\_rotor} = \sqrt{d^2 + l_{m\_rotor}^2} = 0.329m \qquad (6.6)$$

and finally ends up in a total running length for the entire stage of

$$l_{stage} = l_{m\_rotor} + l_{Stator} = 0.482m. \qquad (6.7)$$

From here, we can derive the velocity the disturbances have to travel in order to cross the flow domain in the corresponding time by

$$u = \frac{0.482m}{1.298 \cdot 10^{-03}s} = 371\frac{m}{s}. \qquad (6.8)$$

This is very close to the previously supposed average speed of sound and by considering all made assumptions and simplifications the presumption that these disturbances arise from reflection at the flow domain boundaries and propagate with the speed of sound can hereby be regarded as confirmed. To minimize the influence of these reflections one could either have the simulations run for more iterations or extend the flow domain downstream of the original outlet and/or upstream of the original inlet being aware that both actions would result in higher computational costs.

6 Unsteady Investigations

6.2 Results Overview

Figure 6.4 compares the isentropic efficiency versus normalized mass flow. The continuous lines represent the steady calculations where black diamonds stand for the original stator design and blue squares for the optimized stator design. Of course, it would have been too costly to run an unsteady simulation for each point of the steady speed lines. Thus, only a few points were selected to be run in unsteady mode among which are the the the two points that had been taken for the optimizations. Green squares show the time-averaged values for the original design, red triangles the time-averaged values for the optimized design, respectively.

Apparently, the plots show a very good match between unsteady and steady simulations at design conditions. The differences are rather small and within a good tolerance. However, the discrepancies are much higher near stall. In order to obtain an order of magnitude of these differences Figure 6.5 zooms in on the characteristics at off-design. Both the original and the optimized designs feature a higher efficiency in unsteady mode compared to the corresponding steady simulations. This is a quite common observation when comparing unsteady and steady simulations of a compressor speedline [17] and has mainly two reasons. Firstly, as already described in chapter 4.1, a mixing plane approach was used to model the rotor-stator-interface. For this technique, all flow variables at the interface are averaged in circumferential direction for the rotor outlet as well for the stator inlet. These mixing processes are non-isentropic and therefore artificial sources of loss. Yet, much more serious is the fact, that mixing planes are non-physical and cut through the distributions and gradients of the flow variables. This is particularly grave for compressors with HPA (high-pressure airfoils) where the main load and the highest gradients are located in the front part of the profile as shown by Eulitz [17]. The same holds true for ordinary airfoils when the load is increased towards stall as in this case presented here. Then, these distributions and gradients may reach far into the upstream row when the gap between two adjacent rows tends to be small. Referring back to the investigated compressor, the optimized stator design offers a 0.4% higher efficiency in unsteady mode compared to the steady simulation (green arrow in Figure 6.5) which is a reasonable deviation. However, the difference for the datum stator design is enormous with 1% (blue arrow). We usually expect that unsteady simulations represent a more sophisticated model of the real flow problem. Then, the gain in efficiency we would observe in reality would be considerably smaller than the one, obtained from the steady optimization (black arrow: 1.7% steady-steady ; red arrow: 1.1% unsteady-unsteady). This demands the question whether the optimized geometry would have to look different after all to result in more or less the same efficiency increase in unsteady mode compared to the steady difference or is the impact of the different model approaches so high on the original design with its enormous separation areas that these differences are unavoidable? A method to evaluate which part of the compressor is affected most by switching to unsteady mode is to compare the loss production along the axial extension of the compressor. For this purpose, the entropy flux given by

$$\dot{S} = \int_A s \underbrace{\varrho \vec{u} \vec{n} \mathrm{d}A}_{\dot{m}} \tag{6.9}$$

is plotted in Figure 6.6. The entropy flux represents the specific entropy multiplied by the mass flow. The mass flow is included to avoid any kind of mass deficits which are known to occur between inlet and outlet of the flow domain in unsteady mode (e.g. across the rotor-stator-interface).

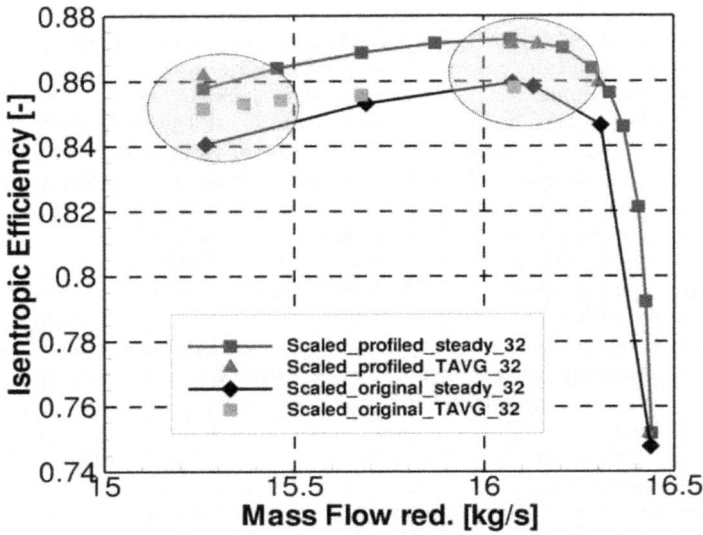

Figure 6.4: Comparison of characteristic lines in steady and unsteady mode

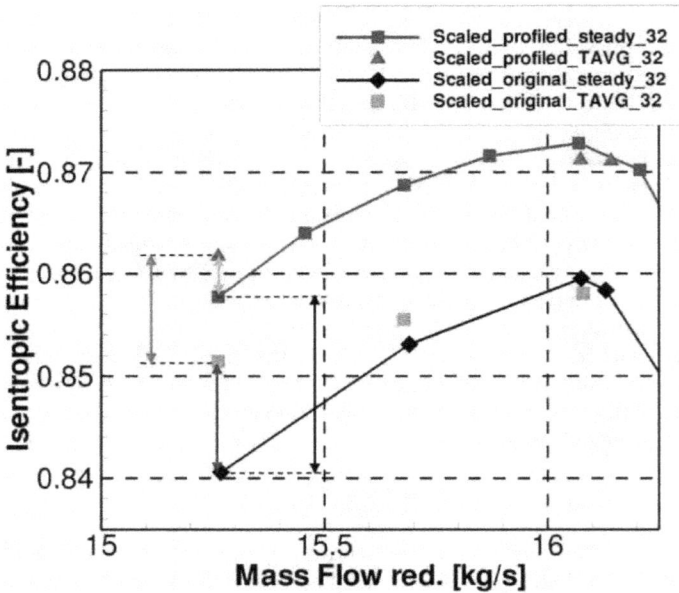

Figure 6.5: Comparison of characteristic lines in steady and unsteady mode at off-design

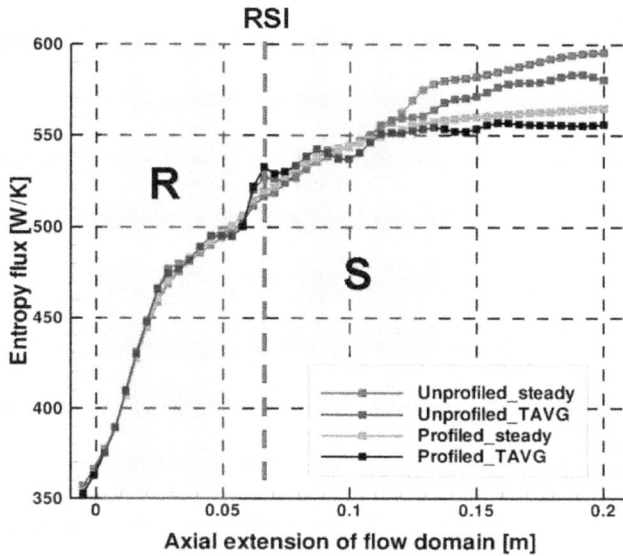

Figure 6.6: Loss production along axial direction of the compressor - Evaluation by entropy flux

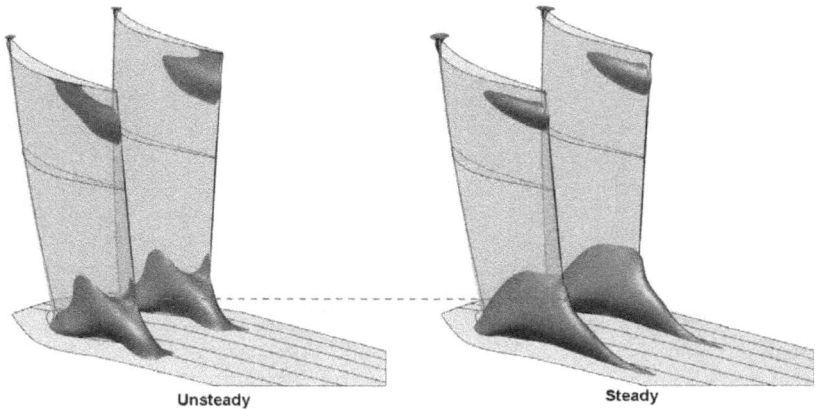

Figure 6.7: Spacial extension of regions of reverse flow in the original axisymmetric stator blade row - Comparison between unsteady and steady mode

According to Figure 6.6. The rotor row of the compressor is not affected in unsteady mode compared to the steady solutions. This is confirmed in numbers by the rotor-only efficiency. In steady state, the original compressor has a rotor efficiency of 88.6% which is only slightly lower at 88.5% for the TAVG value in unsteady mode. Against expectation, both the original (blue line) and the optimized stator designs (black line) do not show increased losses at the rotor-stator interface for steady state. However, the characteristics of the loss production reveal that the aft part of the stator is heavily influenced by the unsteady approach being much more pronounced for the datum design. This leads to the suggestion that the additionally considered phenomena in unsteady mode, in particular affect the separations as the differences are rather small for the optimized design. Therefore, this may be an explanation for the discrepancy at off-design for the datum design.

To substantiate this assumption, Figure 6.7 shows a comparison of the size of the regions with reverse flow for the original stator design for both approaches. The red arrow provides an orientation for the radial extension. When switching to unsteady mode, the separation bubble in the hub region is largely reduced along the last 20% chord length as well as the tongue that reaches far downstream of the stator trailing edge for the steady approach. The dimension of the latter feature is emphasized by the streaklines distribution on the hub surface in Figure 6.8a and 6.8b. The area of streaklines which is highlighted by the red circle in Figure 6.8b indicates that the hub corner continues to exist downstream of the stator for almost one chord length in steady mode. The unsteady simulation in Figure 6.8a shows a different streaklines topology at the stator trailing edge which is more restricted to the blade channel leading to a considerably higher level of static pressure further downstream.

In contrast, the separation area at around 50% chord length has only be changed in form but not in size which is in keeping with the conclusions from the loss development in Figure 6.6 that there are hardly any differences in the first 50% of the stator. In Figure 6.7 the separation area near the casing seems to change only in form from steady to unsteady mode. The radial extension appears to be slightly bigger, in return the bubble is a bit thinner in unsteady mode.

Figures 6.9a and 6.9b display plots of the specific entropy at the trailing edge and aim to impart a better impression of the separation in the pitch wise and the radial directions. Near hub the findings are in agrrement with the observations from Figure 6.7. The unsteady simulation in Figure 6.9b features a much thinner distribution of high entropy in the radial direction according to the spanwise reduced corner stall in the aft part of the blade channel. However, the profiles near shroud only meet the previous observation in so far that the areas of higher entropy have a larger radial extension for the unsteady case and almost spread until mid-span. Inconsistent to Figure 6.7 is the fact that the unsteady simulation results in a much larger distribution of higher entropy in the pitchwise direction although the separation seems to get smaller. Neither does the unsteady distribution show a valley of low entropy between the two peaks located at the top of the stator wake. Furthermore, the steady distribution is characterized by two peak areas of high entropy which are not only greater in terms of their absolute values than in the unsteady mode but also in their size. This is some kind of surprising since the global size of the blade stall near casing was estimated to be around the same for both cases. Thus, it is ambiguous to conclude which approach shows a more beneficial performance in this area. Nonetheless, the most unexpected fact in this context is the large pitchwise expansion of high entropy for the unsteady solution, the separation being not bigger in comparison. An explanation may be found in a changed velocity distribution due to the altered blockage which is influenced by the hub-corner stall. However, this could not completely be clarified within these investigations.

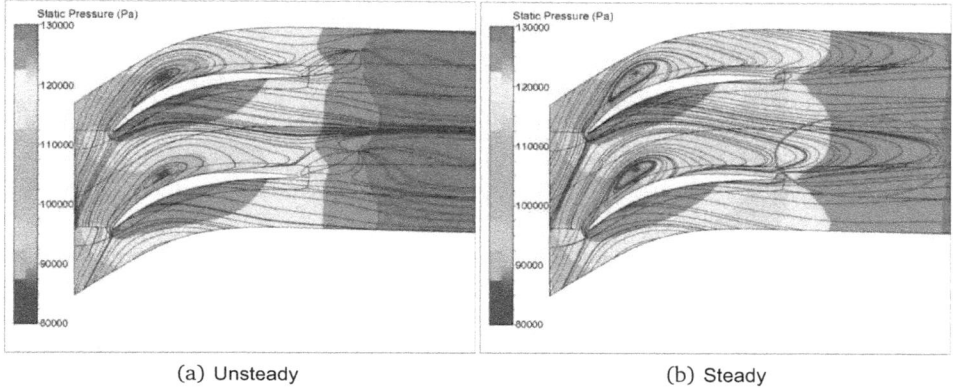

(a) Unsteady   (b) Steady

Figure 6.8: Streaklines and static pressure distribution on stator hub end wall for datum stator design

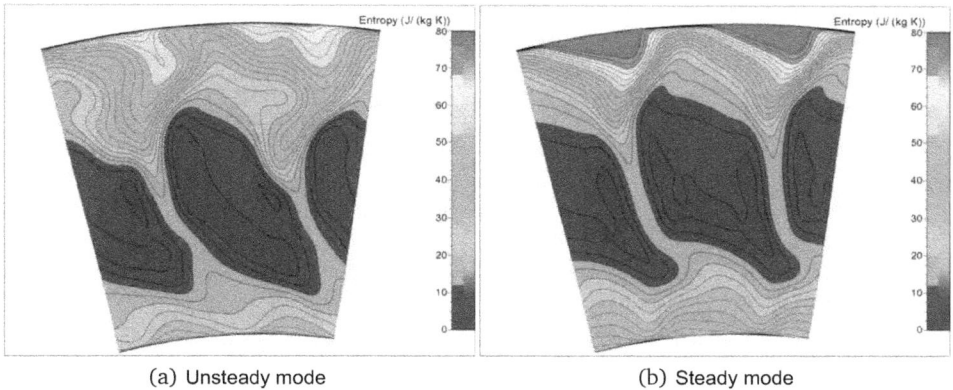

(a) Unsteady mode   (b) Steady mode

Figure 6.9: Contour plots of entropy at stator exit for datum stator design

Figure 6.10 compares the characteristics of the blade stall for the optimized stator design for both approaches. Again, the red arrow provides an orientation for the radial extension. In contrast to the original design, the entire pattern and size are much more similar to each other which also explains the smaller discrepancy in efficiency between steady and unsteady mode for the optimized geometry. Nonetheless, there are still little differences. The red arrow indicates that the blade stall in the unsteady case is of a larger spanwise expansion at the trailing edge. On the other side, we can observe that the flow lifts from the suction side right downstream of the leading edge for the steady simulation which is highlighted by the transparent ellipse. This behavior is less distinct when switching to unsteady mode. For the left blade of the time-averaged solution in Figure 6.10 it even seems that there is only a small core of reverse flow which immediately reattaches and only turns into blade stall further downstream along the aft

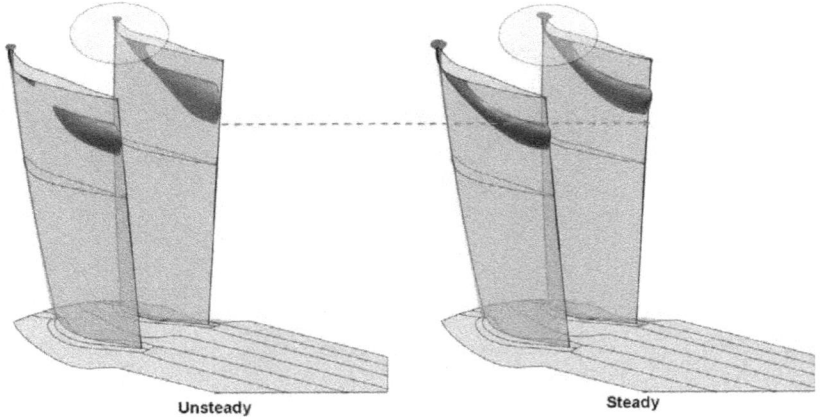

Unsteady   Steady

Figure 6.10: Spacial extension of regions of reverse flow in the optimized non-axisymmetric stator blade row - Comparison between unsteady and steady mode

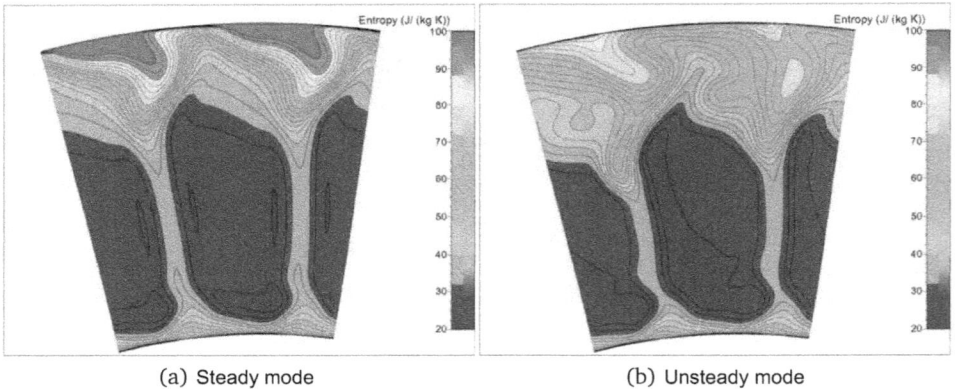

(a) Steady mode   (b) Unsteady mode

Figure 6.11: Contour plots of entropy at stator exit for optimized stator design

50% chord when the blade load is to high for the flow to withstand. In the optimized case, these findings are consistently assured by the contour plots of the specific entropy at the trailing edge in Figure 6.11a and 6.11b. In agreement to the larger spanwise extension of the blade stall, Figure 6.11b reveals a much larger area of a moderate level of entropy whereas the steady simulation leads to much more concentrated pattern with a peak of high entropy in immediate vicinity to the casing. The latter observation is due to the earlier developed blade stall which already covers the tip area of the blade, the produced entropy of which is then transported further downstream.

6.3 Analysis of Discrepancy between Steady and Unsteady Performance of the Original Design

As the differences in performance between the steady and unsteady approach are observed to be rather high for the original stator design (remember a delta in efficiency of 1%), the question must be addressed where these enormous discrepancies come from. Only a satisfying explanation of this issue would still justify the fact, that the compressor has been optimized in steady mode. The findings we have gained so far lead us to the assumption that the different treatments of areas of reverse flow seem to be the key to the explanation. To approach this assumption we will first have a look to an issue from turbine aerodynamics. For low-pressure turbines it is well-known that they may show a separation area on the late suction side when analyzed by steady CFD simulations although this separation can not be observed in rig testing. When switching to unsteady CFD, this stall phenomenon disappears. The reason for that behavior lies in the transitional nature of the boundary layer in low-pressure turbines. The Reynolds numbers tend to be in an order that the boundary layer would normally be laminar. However, additional turbulence is introduced by the rotor wake which leads to an improved momentum transfer in the boundary layer. The boundary layer becomes turbulent and the separation is therewith suppressed. When approaching such a turbine configuration with steady CFD the rotor wake is filtered at the rotor-stator-interface and the lack of the additional turbulence leads to a stable separation on the suction side. When switching to unsteady mode, the rotor wake may leave the rotor domain and enter the vane avoiding the separation. As an example for a

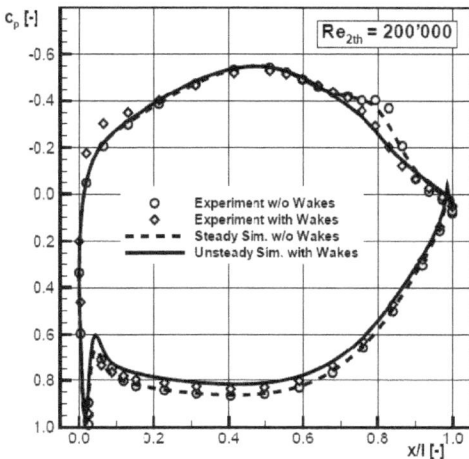

Figure 6.12: Inluence of the rotor wake on the transition effect in a low-pressure turbine illustrated by the profile pressure distribution, from Schwarze et al. [85]

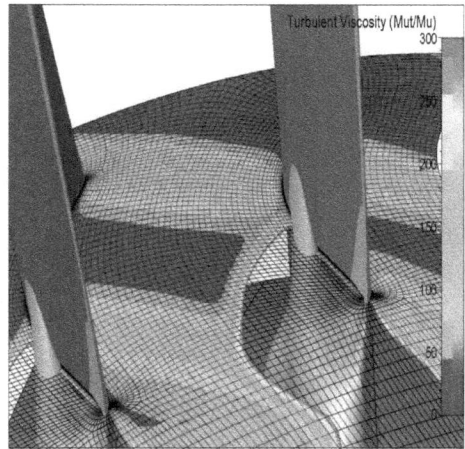

Figure 6.13: Turbulence entrainment by rotor wake and turbulence production by separation phenomena at an arbitrary time step

recently published investigation on this topic, Figure 6.12 shows the influence of periodically unsteady wakes on the boundary layer development and loss behavior of a highly loaded LP turbine cascade at a Reynolds number of 200,000 [85].

Being aware that the boundary layer in the investigated stator row is fully turbulent at a Reynolds number of 1,000,000 and we therefore do not deal with a transitional effect the author is confident that the rotor wake may, nonetheless, have an influence on the extension of the separation by introducing additional turbulence into the stator row. This assumption is supported by the work of Gbadebo et al. [24] who carried out investigations on 3D separation phenomena in axial compressors. He suggested, although not completely proven, that the size of the separated region normal to the blade surface was believed to be sensitive to the turbulent entrainment process at the outer edge of the boundary layer, no matter if the boundary layer was already of turbulent nature.

An adequate parameter to visualize the transport of turbulence is the turbulent viscosity. Figure 6.13 shows the turbulence entrainment by the rotor wake at the tip of the corner stall at around 30% span for an arbitrarily chosen time step for the datum stator design near stall. Unfortunately, this also discloses the trouble in terms of interpretation we get with this kind of illustration. The onset of stall itself leads to a huge additional turbulence production which makes it impossible to distinguish between the turbulence coming from the rotor wake and the turbulence produced by the separation. Of special interest are the areas in the immediate vicinity of the corner stall. Therefore, one has to construct a detour. Since the hub-corner stall is the dominating element and more or less stationary, it is apparent both in the time-averaged solution and for each time step and therefore also its resulting turbulence. In contrast, the rotor wake can only be detected when looking at the results of the individual time step. In the time-averaged solution file, the wake is evened out and no longer visible. Hence, the creation of new solution files in the form of

$$cgns_{filtered} = cgns_{t_i} - cgns_{TAVG} \qquad (6.10)$$

was expected to offer valuable clues to the influence of the rotor wake by filtering out the turbulence with origin from the separation and isolating additionally introduced turbulence by the rotor wake.

Figures 6.14 to 6.16 show the calculated delta in turbulent viscosity referred to the time-averaged mean for some selected time steps for the original stator design at off-design. In order to isolate the influence of the rotor wake only positive values are displayed. Figure 6.14 represents the immediate instant before the rotor wake impinges on the front edge of the separation bubble. The radial limit of the separation bubble towards mid-span is horizontally aligned along almost the entire chord until it drops down at the trailing edge. One time step later, the rotor wake hits the separation in Figure 6.15, the interaction of which instantaneously produces turbulence. This effect is illustrated by the region of high deltas in turbulent viscosity which surrounds the separation bubble. For purpose of a clearer visualization, the peak values are left out by setting an upper bound for the parameter. The radial extension of the corner stall drops directly downstream of the position of interaction but increases again in the spanwise direction at the trailing edge. Overall, less volume is covered by reverse flow than for the previous time step. When we continue to an instant several time steps later in Figure 6.16, we can observe that the highly turbulent rotor wake is about to leave the blade channel. At this time the radial extension of the separation bubble is at its minimum at the trailing edge, the boundary layer receiving a higher transfer of momentum.

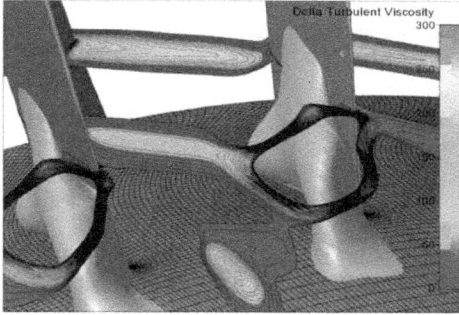

Figure 6.14: Rotor wake before impact on separation bubble, off-design

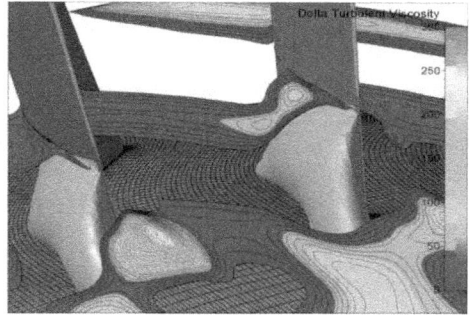

Figure 6.17: Rotor wake before impact on separation bubble, design conditions

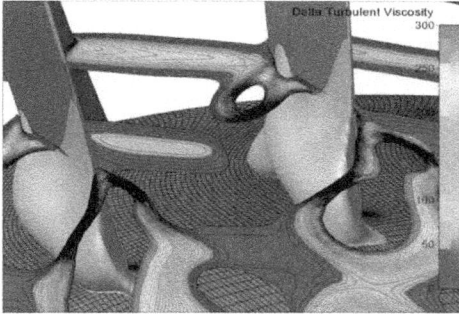

Figure 6.15: Rotor wake impacts the separation bubble, off-design

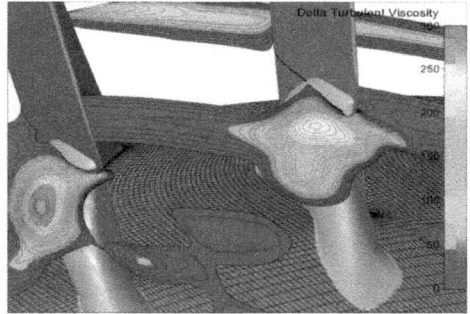

Figure 6.18: Rotor wake impacts the separation bubble, design conditions

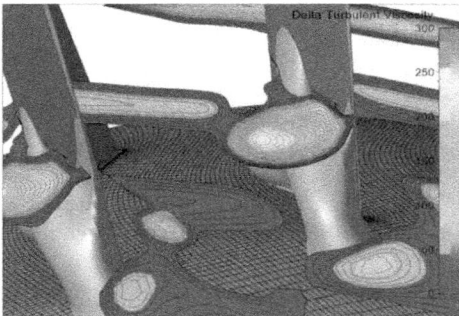

Figure 6.16: Rotor wake leaves the blade passage, off-design

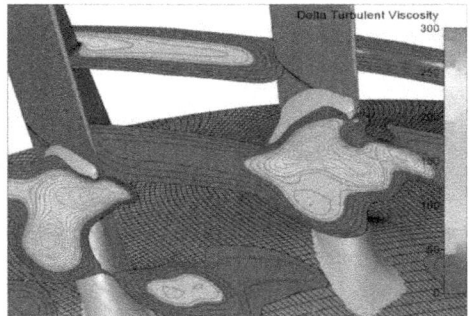

Figure 6.19: Rotor wake leaves the blade passage, design conditions

In contrast, the part of the corner stall towards the leading edge has increased close to its maximum again. However, the radial extension of the hub corner stall is much smaller concerning the time-mean interpretation over one entire passing rotor passage due to the impact of the rotor wake which is filtered out in steady mode.

Of course, one may call this explanation into question as such a big difference in performance and efficiency between the steady and unsteady simulations can only be observed at off-design but not at peak-efficiency although the stator should also see the influence of the rotor wake in the latter case in unsteady mode. To clarify this, Figures 6.17 to 6.19 display the delta in turbulent viscosity for some selected time steps for the original stator configuration at design conditions. Figure 6.17 shows the moment before the rotor wake impinges on the hub-corner stall again. The next time step in Figure 6.18 reveals that the interaction also leads to additional turbulence production at design conditions. However, the separation area proves to be much more compact and stable than at off-design due to the lower blade load which is why the additional turbulence of the rotor wake does not have a big influence on the form and size of the corner stall. For completeness, the time step when the rotor wake leaves the blade channel is shown in Figure 6.19 which accords to the just described observations. Therefore, the rotor wake is assumed only to have a large influence on the form and size of the separation if the separation is characterized by a high instability level due to an increased blade load. Then, small changes in momentum transfer and turbulence entrainment can force the area of reverse flow to flip.

Although this analysis provides a coherent explanation for the discrepancies between the steady and the time-averaged performances, it must be recognized that RANS methods will be subject to error in determining the effective thickness and blockage of a 3D separation due to their limitations in turbulence modeling in turbomachinery, even though the surface patterns are well predicted.

For completeness, Figures for all time steps of one passing rotor passage are listed in the appendix for both described operating points.

6.4 Concluding Assessment of the Steady Optimization

In this final part of the result presentation, a concluding evaluation of the steady optimization process for this compressor configuration will be given. The aim is to answer the question whether unsteady phenomena which could not be captured within the optimization had any negative influence on the flow characteristics of the optimized design. This would involve that the optimized design would have to look different if unsteady effects were considered during the optimization. This is of particular interest when it comes to the industrial application of optimization schemes due to the request for short design cycles. Although the presented optimization approach only requires a comparatively small number of real CFD simulations, a number of 20-30 calculations carried out in unsteady mode would still contradict the mentioned time constraint in the industrial environment.

The first parameters to look at are the efficiency characteristics from Figures 6.4 and 6.5. It shall be maintained that both the original and optimized designs offer a better performance in unsteady mode compared to the steady solutions at off-design. This trend is a major observation although the discrepancies between steady and unsteady efficiency are far from being in the same range due to the different sizes of the corresponding separations. However, this means that

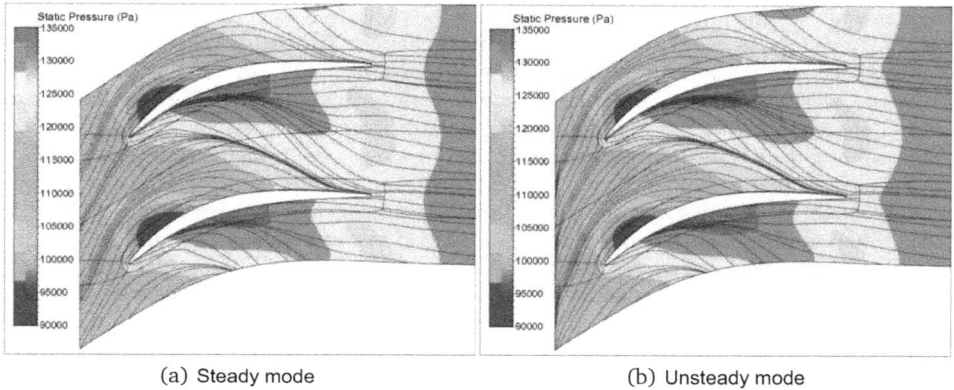

(a) Steady mode   (b) Unsteady mode

Figure 6.20: Streaklines and static pressure distribution on stator hub end wall for optimized stator design

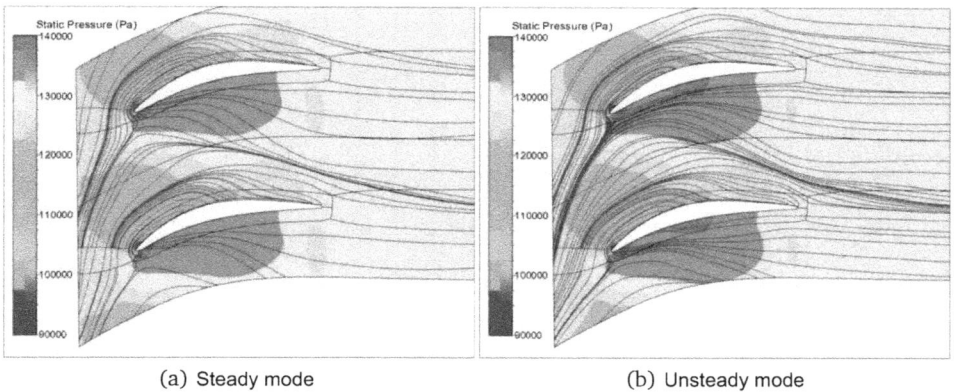

(a) Steady mode   (b) Unsteady mode

Figure 6.21: Streaklines and static pressure distribution on stator casing end wall for optimized stator design

the approaches and the used models are consistent. In order to obtain reasonable information whether the optimized design would have to look different when considering unsteady effects it is not sufficient only to rely on integral values but a detailed comparison of some relevant flow parameters must be involved.

In chapters 4 and 5, plots of the static pressure distribution on the end walls have proven to provide conclusive explanations for the observed improvements. Additionally, the static pressure distributions are directly connected to the forms of the non-axisymmetric end wall contours. Therefore, a comparison of the static pressure distribution on the hub surface together with streaklines near stall is given in Figure 6.20 for the optimized stator design. As we can observe, the static pressure distributions on the end walls correspond to each other and also the streaklines show an identical behavior. In fact, the unsteady mode even shows a slightly higher level of static pressure (red area) close to the pressure side. This tongue (orange zone) of high pressure also reaches further into the blade channel. This means that the found optimal hub end wall design has an even more positive effect on the performance in unsteady mode.

To conclude, Figure 6.21 shows the static pressure distribution together with streaklines on the shroud surface at off-design conditions for both the steady and unsteady simulations. These plots confirm the just described observations offering also a slightly higher static pressure close to the pressure side.

This finally leads to the judgment that there is no need that the contours would have to look different if the optimization process was carried out in unsteady mode as the identical positive effects of the end wall modifications can be observed for both the steady and unsteady approaches. For this configuration, unsteady phenomena can therewith be neglected during the optimization process.

7 Conclusions

7.1 Summary

The emphasis of this work has been to analyze the steady and unsteady performance of a transonic compressor stage with non-axisymmetric end walls where the profiled end wall serves to control the addressed flow inhomogeneities such as separation phenomena and secondary flow. After an introduction to the field of end wall contouring an efficient method for constructing non-axisymmetric end walls has been presented and applied to the stator and rotor end walls of Configuration I of the Darmstadt Transonic Compressor. The tool for this purpose consists of a fully-automated multi-objective optimizer connected to a steady 3D-RANS flow solver. The goal has been to analyze how effective such a design tool can work on such a challenging task and to derive first design rules and compare the differences and features in common to the experience made by turbine researchers.

The study on the stator row has demonstrated the possibility of suppressing separation phenomena in a stator row with 2D blading by applying profiled end walls. The new hub design led to an increase in efficiency of 1.8% due to the suppression of the hub-corner stall. However, this was accompanied by an increased area of reverse flow at the casing, which was even more distinct at off-design conditions near stall. The numerical surge limit of the datum axisymmetric design could no longer be reached and was then determined by the new separation close to the stator casing. A subsequent optimization of the shroud end wall was carried out using the improved profiled hub as the initial design. An operating point near stall with a strongly developed separation was chosen for this purpose. The second optimization resulted in a further improvement in the characteristic speed line over the entire off-design region. Although the shroud contour was designed at off-design conditions, the optimization gained an additional 0.03% in efficiency for the design point. The lower surge limit of the datum design could also be reached again, even at higher efficiency and pressure ratios. All over, the exit flow field's whirl angle distribution was made more uniform which is generally desirable for the inlet conditions of a downstream stage. The changed static pressure distribution, resulted in the migration of the loss cores from the end wall onto to the airfoil suction surface. This was identified as the main mechanism responsible for the obtained results. According to the study, it makes sense to design the two end walls at different operating points to allow the optimizer to exploit its full potential as the separation region in the tip area is not sufficiently distinct at the design point after the hub optimization. Most of the secondary loss reduction (according to the used definition) within the stator hub optimization could be ascribed to the suppressed hub-corner stall, showing again the difficulty of defining a primary flow field. In contrast, secondary loss could not be reduced within the stator shroud optimization, even though the distribution considerably changed. This is not surprising since secondary flow, following the classical theory, has not been reduced which was expected to result in less overturning in the rear part of the blade channel. The application of non-axisymmetric end walls rather increased the classical secondary flow by enhancing the cross flow from the pressure to the suction side. However, this contradictory phenomenon helped to avoid separation. This leads to the following assumptions: either the used definition of the secondary flow is not really appropriate and an alternative one is required or a rise in efficiency by suppressing separation in transonic compressors is not necessarily connected to a reduction in SKE. Separation phenomena in modern multi-stage compressors are kept under control by 3D blade shaping. The potential for the application of non-axisymmetric end walls with 3D airfoils therefore remains to be shown. Moreover, the benefits would then have to be large enough to justify the additional manufacturing costs. Nonetheless, reducing the axial chord length and smoothing the effect of the resultant higher blade load by end wall profiling may be conceivable. This in turn would lead to a reduction in engine size and weight. Possible applications will be addressed in the outlook section.

Subsequently, the same process chain was applied to the rotor hub end wall of Configuration I of the Darmstadt Transonic Compressor. Several optimization strategies involving different objective functions to be minimized and the corresponding performances were compared. The parameters considered within the optimization process were isentropic stage efficiency, pressure loss in the rotor, throat area and secondary kinetic energy (SKE). For the best design, the isentropic efficiency could be raised by 0.07%, SKE at the rotor exit was reduced while the total pressure ratio of the stage remained constant. A mechanism similar to the principle of lean has been observed for the profiled rotor, an effect on which literature had not been reported before to the knowledge of the author. The non-axisymmetric end walls produced blockages and reduced the loss production in the outer regions (>5% channel height) at the expense of an increase in near hub end wall losses. This effect complements the observations on the impacts on the migration of corner vortices and on the modifications of the shock pattern in transonic compressor rows which could be derrived by other researches. This mechanism is also quite different from the flow mechanism identified to the improvements of the contoured stator row. Comparing the different design cases, some first design rules for the construction of non-axisymmetric end walls in compressors could be derived, consisting of: a raised area in the middle of the rear part of the blade channel, a small dip at the suction side of the leading edge and a strong depression at the pressure side. Furthermore, a new method based on Euler walls has been introduced to visualize the evaluation of secondary flows. This method clearly shows how the characteristics of secondary flow can be positively influenced by using non-axisymmetric end walls. Since this seems to be an adequate method, it could be used within the optimization process although this will be very time-consuming due to the second CFD simulation required.

As a further step, the obtained optimized geometries have been investigated in unsteady mode. The simulations show a very good agreement between steady and TAVG results at design condition but rather large discrepancies at off-design. Here, the gain in efficiency has been found to be 1.7% in steady compared to 1.1% for the TAVG values. However, it is crucial that both the original and optimized designs show a better performance in unsteady mode compared to the steady approach which proofs the models to be consistent. The different behavior has been observed to start in the aft part of stator row which has led to the assumption that the higher deviation for the original design are due to influence of the rotor wake similar to LP Turbines, although not of transitional nature but most likely as a result of the different momentum transfer in the boundary layer. This has been confirmed by the constructing artifical solution files where the turbulence production of the corner stall is filtered in order to isolate the effect of the rotor wake.

Therefore, the mismatch between the deltas in efficiency gained through the steady optimization and the deltas in unsteady mode are due to the different approches and is not connected to the obtained end wall geometries. This is supported by the fact that static pressure distributions on end walls as the key factor correspond to each other in steady and unsteady mode. Hence, there is no need that the contour would have to look different if unsteady effects had been considered during the optimization as the positive effect can be observed for both steady and unsteady mode. This leads to the final conclusion that, for this compressor configuration, unsteady phenomena can be neglected during the optimization process.

7.2 Outlook

In this thesis it has been demonstrated that profiled end walls can be used to affect the end wall flow field in a significant way and to control separation phenomena in the absence of 3D airfoils. For modern multi-stage compressors secondary flows and the associated losses are rather small while any corner stall in the end wall regions is kept under control by the application of sweep and dihedral when blade load is increased towards stall. Therefore, it remains to be shown how non-axisymmetric end walls might be combined with 3D airfoils. Most likely, the first opportunities for the successful application might be to improve compressor performance where 3D airfoils cannot be used due to mechanical or other design constraints or in a retrospective integration. Nonetheless, the observed findings and results within this thesis suggest several possible applications of non-axisymmetric end walls for compressor configurations. Hence, this work may be continued by the following topics:

- As stated in the introduction, CFD always relies on experimental data in order to confirm and to further develop the employed solvers and models. This is in particular true for turbulence modelling, which is still a crucial issue. In the context of this thesis, it is also important to check the capability of the optimizer to predict designs with a better performance. Thus, it would be reasonable to have the optimized stator design manufactured and rig-tested. The deltas received out of the optimization process should be large enough to be also noticeable in the experiments even if measurement uncertainties are considered. Moreover, the manufacturing of a stator is comparatively cheap in contrast to a transonic rotor. The observed deltas for the rotor optimization are rather small and may lie within the measurements tolerances. Therefore, an experimental investigation of the optimized stator design is definitely to be prioritized.

- From turbine research it is known, that the development of secondary flow can either be influenced by the transverse static pressure gradient or the formation of the horseshoe vortex. The latter effect, in turn, can be influenced by either applying bulbs or fillet radii. Fillet radii have not been included into the optimization process since the used grid generator has not been able to handle profiled end walls and fillets in parallel at the time of the optimization. The latest version of the grid generator, however, is capable to handle both features at once. Therefore, it would be worthwhile to include the fillet radii into the optimization process. The presented new method to visualize secondary flow structures and in particular the formation of the horseshoe vortex makes the author confident that the influence of the fillets on the flow field may be better understood.

- Although corner stall is suppressed by 3D airfoils in modern compressors, the application of profiled end walls may show potential to reduce the blade count in a well-designed compressor while maintaining the original stage loading. First promising investigations

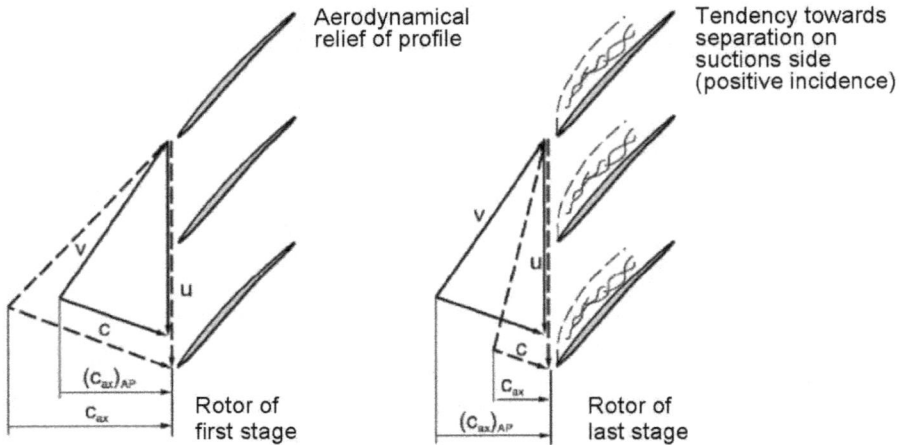

Figure 7.1: Altered velocity triangles at the rotor inlet of the first and the last stage in a multi-stage compressor configuration for rotational speed above design speed, taken from Bräunling [6]

have been carried out by Heinichen et al. [43]. They reduced the number of blades in a well-designed compressor stator which contained 3D design features by around 18%. This resulted in the development of a pronounced flow separation zone close to the stator casing. Surprisingly, this corner stall could not be repaired by adjusting the 3D design features of the blading but only by the retrospective application of profiled end walls. These findings show a very promising potential for profiled end walls to reduce weight and length for jet engines. Concerning all rows which are not manufactured as an integral component (BLISK-Bladed Disk), this would also lead to benefits in terms of cost savings. The latter aspect makes this technique especially interesting for heavy-duty gas turbines where the costs for a single blade may be much larger due the larger size of the blades.

- The fact, that the suppression of separation zones has been identified as the main flow feature that can be controlled by end wall contouring results in a promising application in multi-stage axial compressors. Figure 7.1 shows the rotors of the first and the last stages of an axial compressor with the corrsponding velocity triangles for a blade speed above design speed. According to the explanations in Bräunling [6], the mass flow and the total pressure ratio increase if a compressor is run along its operational line towards higher rotational speeds, i.e. for $n > n_{DP}$. A higher compression takes place which the airfoils are not optimized for. This is e.g. the case during take-off and climb when a jet engine is operated at above its design point. At the compressor inlet, the increased mass flow leads to an increased axial velocity in keeping with the continuity equation as the density remains unchanged at constant ambient conditions. The front stage therefore experiences a negative incidence together with an aerodynamic relief of the profile. In contrast, the higher compression leads to a smaller axial velocity at the inlet of the last stage. There, the profile experiences a positive incidence with a high tendency towards separation on the

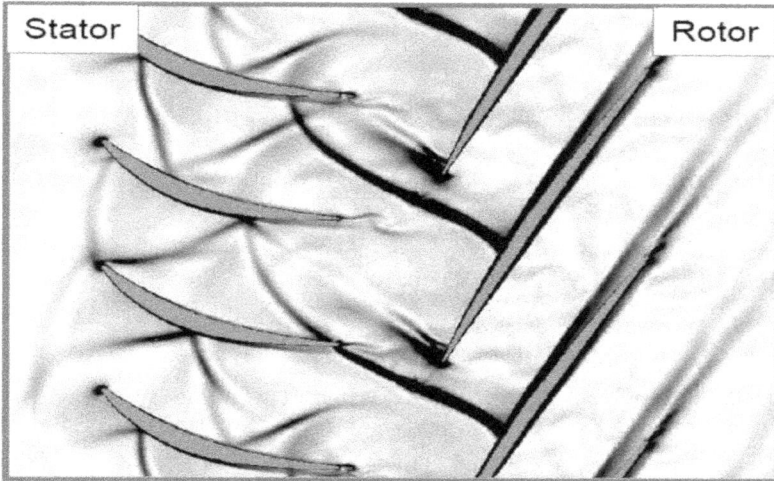

Figure 7.2: Shock pattern of a transonic compressor rotor travelling into the upstream stator row, taken from Eulitz [17]

suction side. The fatal issue is that if under these conditions the last stage begins to stall, the aerodynamically relieved front stages will no longer be able to support the pressure rise. Therefore, the entire compressor will stall and even be destroyed which is absolutely unacceptable due to safety requirements. These separations at overload may be suppressed by profiled end walls which would yield to a possible increase of the operating range by stabilizing the hub-critical aft stages in multi-stage configurations.

- For Configuration I of the Darmstadt Transonic Compressor, it has been shown that unsteady phenomena could be neglected during the optimization process. However, this might completely changed if e.g. an IGV is included in the compressor arrangement or if more than one transonic stage is present. Shock pattern of the transonic rotor row may then travel into the stator of the upstream stage or the IGV as shown in Figure 7.2 and have a major impact on the performance of the corresponding row. The interaction of the shock pattern is filtered at the rotor-stator-interface in steady mode and therefore impossible to be considered during a steady optimization. It would be interesting to investigate the impact of such shock pattern on stall phenomena whether they are amplified or damped. In order to find an optimum design, unsteady effects would have to be considered within the optimization process. At present, unsteady optimization is a matter of research but very costly in terms of CPU-time. Most likely, alternative approaches to deal with the unsteady RANS-equation such as frequency domain techniques will be necessary within the optimization schemes in order to make it affordable for the industrial use in the future.

Bibliography

[1] BARANKIEWICZ W., HATHAWAY M.: *Impact of Variable-Geometry Stator Hub Leakage in a Low Speed Axial Compressor*. ASME Paper GT1998-194, 1998.

[2] BECZ S., MAJEWSKI M.S., LANGSTON L.S.: *Leading Edge Modification Effects on Turbine Cascade Endwall Loss*. ASME GT2003-38898, 2003.

[3] BECZ S., MAJEWSKI M.S., LANGSTON L.S.: *An Experimental Investigation of Contoured Leading Edges For Secondary Flow Loss Reduction*. ASME Paper GT2004-53964, 2004.

[4] BIELA C., MÜLLER M.W., SCHIFFER H.-P. ,ZSCHERP C.: *Unsteady Pressure Measurement in a Single Stage Axial Transonic Compressor near the Stability Limit*. ASME Paper GT2008-50245, 2008.

[5] BLACKBURN J., FRENDT G., GAGNE M., GENEST J.-D., KOHLER T., NOLAN B.: *Performance Enhancements to the Industrial Avon Gas Turbine*. ASME Paper GT2007-28315, 2007.

[6] BRÄUNLING, WILLY J. G.: *Flugzeugtriebwerke*. Springer-Verlag, Berlin, 2004.

[7] BRENNAN G., HARVEY N., ROSE M., FOMISON N., TAYLOR M.: *Improving the Efficiency of the Trent 500 HP Turbine Using Nonaxisymmetric End Walls - Part I: Turbine Design*. Journal of Turbomachinery, Vol. 125 Issue 3:497–504, 2003.

[8] CASEY, M.V.: *Quality Design with ASC's CFD Simulations*. Advanced Scientific Computing LTd. Waterloo, Ontario, Canada, Vol.3, No.3, 1996.

[9] CUMPSTY, N.A.: *Compressor Aerodynamics*. Krieger Publishing Company, 2004.

[10] DEMEULENAERE, A.: *Technical Note on Modeling of Rotor-Stator Interactions*. Numeca International s.a., 2002.

[11] DEMEULENAERE A., HIRSCH CH.: *Application of Multipoint Optimisation to the Design of Turbomachinery Blades*. ASME Paper GT2004-53110, 2004.

[12] DENTON, J.D.: *Loss Mechanisms in Turbomachines*. Journal of Turbomachinery, 115:621–656, 1993.

[13] DENTON J.D., XU L.: *The Exploitation of Three-Dimensional Flow in Turbomachinery Design*. Proc. Instn. Mech. Engrs., 213 Part C:125–137, 1999.

[14] DENTON J.D., XU L.: *The Effects of Lean and Sweep on Transonic Fan Performance*. ASME GT2002-30327, 2002.

[15] DES, GRADUIERTENKOLLEGS 1344 DOKTORANDEN: *Projektplanung zur Untersuchung instationärer Effekte in einem Flugtriebwerk am Beispiel von Windmill Relight*. TU-Darmstadt, 2009.

[16] DORFNER C., NICKE E., VOSS C.: *Axis-Asymetric Profiled Endwall Design using Multiobjective Optimization Linked with 3D RANS-Flow-Simulation.* ASME Paper GT2007-27268, 2007.

[17] EULITZ, F.: *Unsteady Flow Simulations.* VKI Lecture Series 2009-08: Numerical Investigation in Turbomachinery: The State of the Art, 2009.

[18] EYMANN S., REINMÖLLER, NIEHUIS R. , FÖRSTER W., BEVERSDORFF M., GIER J.: *Improving 3D Flow Characteristics in a Multistage LP Turbine by Means of Endwall Contouring and Airfoil Design Modification - Part 1: Design and Experimental Investigation.* ASME GT2002-30352, 2002.

[19] FAVRE, A.: *Equations des gaz turbulents compressible, part 1: formes générales.* Journal de Méchanique 4, pp:361 – 390, 1965.

[20] FAVRE, A..: *Equations des gaz turbulents compressible, part 2: méthode de vitesses moyennes; méthode des vitesse moyennes pondérées par la masse volumique.* Journal de Méchanique 4, pp:361 – 390, 1965.

[21] FRIEDRICHS S., HODSON H.P., DAWES W.N.: *Distribution of Film Cooling Effectiveness on a Turbine Endwall Measured Using the Ammonia Diazo Technique.* Journal of Turbomachinery, Vol. 118:pp. 613–621, 1996.

[22] GALLIMORE S.J., BOLGER J., CUMPSTY N., TAYLOR M., WRIGHT P., PLACE M.: *The Use of Sweep and Dihedral in Multistage Axial Flow Compressor Blading - Part I: University Research and Methods Development.* Journal of Turbomachinery, Vol. 124:pp. 521–532, 2005.

[23] GALLIMORE S.J., BOLGER J., CUMPSTY N., TAYLOR M., WRIGHT P., PLACE M.: *The Use of Sweep and Dihedral in Multistage Axial Flow Compressor Blading - Part II:Low and High-Speed Designs and Test Verifications.* Journal of Turbomachinery, Vol. 124:pp. 533–541, 2005.

[24] GBADEBO A., CUMPSTY N., HYNES T.: *Three-Dimensional Separation in Axial Compressors.* ASME Paper GT2004-53617, 2004.

[25] GERMAIN T., NAGEL M., BAIER R.-D.: *Visualisation and Quantification of Secondary Flows: Application to Turbine Bladings with 3D-Endwalls.* Paper ISAIF8-0098, 2007.

[26] GERMAIN T., NAGEL M., RAAB I., SCHUEPBACH P., ABHARI R.S., ROSE M.: *Improving Efficiency of a High Work Turbine Using Non-Axisymmetric Endwalls Part I: Endwall Design and Performance.* ASME Paper GT2008-50469, 2008.

[27] GOLDBERG, D.E: *Genetic Algorithm.* Addison Wesley, 1994.

[28] GOURDAIN N., LEBOEUF F.: *Unsteady Simulations of an Axial Compressor Stage with Casing and Blade Passive Treatments.* Journal of Turbomachinery, Vol. 131 Issue 2:pp. 021013–1 – 021013–12, 2009.

[29] GREGORY-SMITH D., INGRAM G., JAYARAMAN P., HARVEY N., ROSE M.G.: *Non-Axisymmetric Turbine End Wall Profiling.* Proceedings of the 4th European Conference on Turbomachinery, 2001.

[30] GÜMMER V., WENGER U., KAU H.-P.: *Using Sweep and Dihedral to Control Three-Dimensional Flow in Transonic Stators of Axial Compressors.* ASME Paper GT2000-0491, 2000.

[31] GÜMMER V., GOLLER M., SWOBODA M.: *Numerical Investigation of Endwall Boundary Layer Removal on Highly-Loaded Compressor Blade Rows.* ASME Paper GT2005-68699, 2005.

[32] GUSTAFSON R., MAHMOOD G., ACHARYA S.: *Aerodynamic Measurements in a Linear Turbine Blade Passage with Three-Dimensional Endwall Contouring.* ASME Paper GT2007-28073, 2007.

[33] HAH C., BERGNER J., SCHIFFER H.-P.: *Short Length Rotating Stall Inception in a Transonic Axial Compressor - Criteria and Mechanisms.* ASME Paper GT2006-90045, 2006.

[34] HARLAND J., GREGORY-SMITH D., HARVEY N., ROSE M.: *Nonaxisymmetric Turbine End Wall Design: Part II - Experimental Validation.* Journal of Turbomachinery, Vol. 122 Issue 2:286–293, 2000.

[35] HARLAND J., GREGORY-SMITH D., ROSE M.: *Non-axisymmetric End Wall Profiling in a Turbine Rotor Blade.* ASME GT1998-525, 1998.

[36] HARRISON, S.: *The Influence of Blade Lean on Turbine Losses.* Journal of Turbomachinery, Vol. 114 Issue 1:pp. 184–190, 1992.

[37] HARVEY, N. *personal communication*, 2009.

[38] HARVEY, N.W.: *Some Effects of Non-Axisymmetric End Wall Profiling on Axial Flow Compressor Aerodynamics. Part I: Linear Cascade Investigation.* ASME Paper GT2008-50990, 2008.

[39] HARVEY N.W., OFFORD T.P.: *Some Effects of Non-Axisymmetric End Wall Profiling on Axial Flow Compressor Aerodynamics. Part II: Multi-Stage HPC CFD Study.* ASME Paper GT2008-50991, 2008.

[40] HARVEY N.W., NEWMAN D.A., ROSE M.: *Improving Turbine Efficiency Using Non-Axisymmetric End Walls: Validation in the Multi-Row Environment and with Low Aspect Ratio Blading.* ASME Paper GT2002-30337, 2002.

[41] HASELBACH F., SCHIFFER H.-P., HORSMAN H., DRESSEN S., HARVEY N.W., READ S.: *The Application of Ultra-High Lift Blading in the BR715 LP Turbine.* ASME Paper GT2001-0436, 2001.

[42] HAWTHORNE, W.R.: *Secondary Circulation in Fluid Flow.* Proceedings of the Royal Society, A206:pp. 374–387, 2000.

[43] HEINICHEN F., GÜMMER V., SCHIFFER H.-P.: *Extending the Design Parameter Space of a Highly Loaded High Pressure Compressor Stator through Application of Non-Axisymmetric Endwall Contouring.* DGLR-Jahreskongress, 2009.

[44] HERGT C., MEYER R, MÜLLER M.W., ENGEL K.: *Loss Reduction in Compressor Cascades by Means of Passiv Flow Control.* ASME Paper GT2008-50357, 2008.

[45] HILDEBRANDT, T.: *Weiterentwicklung von 3D Navier-Stokes-Strömungsrechenverfahren zur Anwendung in hochbelasteten Verdichter- und Turbinengittern.* Dissertation, Universität der Bundeswehr München, 1998.

[46] HILDEBRANDT T., THIEL P.: *CFD Based Aerodynamic Optimisation of a Turbocharger in Multiple Operating Points.* NAFEMS Paper, 2008.

[47] HIRSCH, C.: *Numerical Computation of Internal and External Flows, Second Edition: The Fundamentals of Computational Fluid Dynamics.* Krieger Publishing Company, 2007.

[48] HIRSCH, CH.: *Numerical Methods in Turbomachinery: Fundamentals.* VKI Lecture Series 2009-08: Numerical Investigation in Turbomachinery: The State of the Art, 2009.

[49] HOEGER M., CARDAMONE P., FOTTNER L.: *Influence of Endwall Contouring on the Transonic Flow in a Compressor Blade.* ASME Paper GT2002-30440, 2002.

[50] HOEGER M., SIEVERS N., LAWERENZ M.: *On the Performance of Compressor Blades with Contoured End Walls.* Proceedings of the 4th European Conference on Turbomachinery, 2001.

[51] HOEGER M., BAIER R.D., MÜLLER R., ENGBER M.: *Impact of a Fillet on Diffusing Vane Endwall Flow Structure.* ISROMAC Paper 2006-057, 2006.

[52] ILIOPOULOU V., LEPOT I., GEUZAINE P.: *Design Optimization of a HP Compressor Blade and its Hub Endwall.* ASME-2008-50293, 2008.

[53] INGRAM G., GREGORY-SMITH D., HARVEY N.: *Investigation of a Novel Secondary Flow Feature in a Turbine Cascade with EndWall Profiling.* ASME Paper GT2004-53589, 2004.

[54] INGRAM G., GREGORY-SMITH D., ROSE M., HARVEY N., BRENNAN G.: *The Effect of End-Wall Profiling on Secondary Flow and Loss Development in a Turbine Cascade.* ASME Paper GT2002-30339, 2002.

[55] JANICKA, J.: *Verbrennungstechnologie in der Energieumwandlung I.* Skriptum zur Vorlesung, TU-Darmstadt - FG Energie- und Kraftwerkstechnik.

[56] KESKIN, A.: *Process Integration and Automated Multi-Objective Optimization Supporting Aerodynamic Compressor Design.* Dissertation, Brandenburgische Technische Universität Cottbus, 2006.

[57] LARRY, K.: *Unsteady Investigations on the Potentional of Non-Axisymmetric Endwalls in a Transonic Compressor.* Technische Universität Darmstadt, 2009.

[58] LEI V.-M., SPAKOVSZKY S., GREITZER E.M.: *A Criterion for Axial Compressor Hub-Corner Stall.* ASME Paper GT2006-91332, 2006.

[59] LEPOT I., ILIOPOULOU V., MANZINI G., SIMON J.-F.: *3D Endwall Profiling Impact on Axial Flow Compressor Aerodynamics.* ISABE-2009-1101, 2009.

[60] LIEBLEIN S., SCHWENK F.C., BRODERICK R. L.: *Diffusion Factor for Estimation of Losses and Limiting Blade Loadings in Axial Flow Compressor Blade Elements.* ANACA RM E53D01, 1953.

[61] MAHMOOD G., ACHARYA S.: *Measured Endwall Flow and Passage Heat Transfer in a Linear Blade Passage with Endwall and Leading Edge Modifications.* ASME Paper GT2007-28179, 2007.

[62] MARCHAL P., SIEVERDING C.H.: *Secondary Flows within Turbomachinery Bladings.* AGARD-CP-214, 1977.

[63] MÜLLER M.W., SCHIFFER H.-P., HAH C.: *Effect of Circumferential Grooves on the Aerodynamic Performance of an Axial Single-Stage Transonic Compressor.* ASME Paper GT2007-27365, 2007.

[64] MÜLLER M.W., BIELA C., SCHIFFER H.-P., HAH C.: *Interaction of Rotor and Casing Treatment in an Axial Single-Stage Transonic Compressor.* ASME Paper GT2008-50135, 2008.

[65] MÜLLER R., SAUER H., VOGELER K.: *Influencing the Secondary Losses in Compressor Cascade by a Leading Edge Bulb Modification at the Endwall.* ASME Paper GT2002-30442, 2002.

[66] MÜLLER R., SAUER H., VOGELER K., HOEGER M.: *Endwall Boundary Layer Control in Compressor Cascades.* ASME Paper GT2004-53433, 2004.

[67] NAGEL M.G., BAIER R.D.: *Experimentally Verified Numerical Optimization of a Three-Dimensional Parameterized Turbine Vane with Nonaxisymmetric End Walls.* Journal of Turbomachinery, Vol. 125:380–387, 2005.

[68] NGUYEN B.Q., SQUIRES K.D.: *A simple Procedure to Reduce Secondary Flow Effect in Turbine Nozzle Guide Vane.* ASME Paper GT2007-28159, 2007.

[69] NIEHUIS R., LÜCKING P., STUBERT B.: *Experimental and Numerical Study on Basic Phenomena of Secondary Flows in Turbines.* AGARD-CPP-469, 1989.

[70] NUMECA, INTERNATIONAL: *User Manual $FINE^{TM}/Design3D$ v3.* 2007.

[71] NUMECA, INTERNATIONAL: *User Manual $FINE^{TM}/Turbo$ v8 Flow Integrated Environment.* 2007.

[72] PIERRET S., DEMEULENAERE A., GOUVERNEUR B., HIRSCH C.: *Designing Turbomachinery Blades with the Function Approximation Concept and the Navier-Stokes Equation.* AIAA 2000-4879, 2000.

[73] PRAISNER T.J., ALLEN-BRADLEY E., GROVER E.A., KNEZEVICI D.C., SJOLANDER S.A.: *Application of Non-Axisymmetric Endwall Contouring to Conventional and High-Lift Turbine Airfoils.* ASME Paper GT2007-27579, 2007.

[74] REISING S., SCHIFFER H.-P., FONT BROSSA J.: *CFD Analysis of Hub-Corner Stall and Secondary Flow in a Transonic Compressor Stage with Non-Axisymmetric End Walls.* Proceedings of the 8th European Conference on Turbomachinery, 2009.

[75] REISING S., SCHIFFER H.-P., HILDEBRANDT T., THIEL P.: *Automated Aerodynamic Optimisation Of a Transonic Compressor Stage by Application of Non-Axisymmetric End Walls.* DGLR-Jahreskongress ID81252, 2008.

[76] REISING S., SCHIFFER H.-P.: *Non-Axisymmetric End Wall Profiling in Transonic Compressors. Part I: Improving the Static Pressure Recovery at Off-Design Conditions by Sequential Hub and Shroud End Wall Profiling.* ASME Paper GT2009-59133, 2009.

[77] REISING S., SCHIFFER H.-P.: *Non-Axisymmetric End Wall Profiling in Transonic Compressors. Part II: Design Study of a Transonic Compressor Rotor Using Non-Axisymmetric End Walls - Optimization Strategies and Performance.* ASME Paper GT2009-59134, 2009.

[78] ROLLS-ROYCE: *The Jet Engineering, 6. Auflage.* Rolls-Royce plc, 2006.

[79] ROSE, M.G.: *Non-Axisymmetric Endwall Profiling in the HP NGV's of an Axial Flow Gas Turbine.* ASME GT1994-249, 1994.

[80] ROSE M., HARVEY N., SEAMAN P., NEWMAN D., MCMANUS D.: *Improving the Efficiency of the Trent 500 HP Turbine Using Nonaxisymmetric End Walls - Part II: Experimental Validation.* ASME GT2001-0505, 2001.

[81] ROTH, M.: *Automatic Extraction of Vortex Core Lines and Other Line-Type Features of Scientific Visualization.* Dissertation, ETH Zürich, 2000.

[82] SAUER H., MÜLLER R., VOGELER K.: *Reduction of Secondary Flow Losses in Turbine Cascades by Leading Edge Modification at the End Wall.* ASME GT2000-0473, 2000.

[83] SCHUEPBACH P., ABHARI R.S., ROSE M., GERMAIN T., GIER J., RAAB I.: *Improving Efficiency of a High Work Turbine Using Non-Axisymmetric Endwalls Part I: Endwall Design and Performance.* ASME Paper GT2008-50469, 2008.

[84] SCHULZE, GISBERT: *Betriebsverhalten eines transsonischen Axialverdichters.* Dissertation, Technische Universität Darmstadt, 1996.

[85] SCHWARZE M., MARTINSTETTER M., NIEHUIS R., KOTZBACHER T.: *Unsteady Numerical Simulation of the Influence of Periodically Unsteady Wakes on Boundary Development and Loss Behavior of a Highly Loaded LP Turbine Cascade.* ETC, 2009.

[86] SMITH L.H., YEH H.: *Sweep and Dihedral Effects in Axial Flow Turbomachinery.* ASME Journal of Basic Engineering, 1963.

[87] SONODA T., HASENJÄGER M., ARIMA T., SENDHOFF B.: *Effect of Endwall Contouring on Performance of Ultra-Low Aspect Ratio Transonic Turbine Inlet Guide Vanes.* ASME GT2007-28210, 2007.

[88] SPALART P. R., ALLMARAS S.R.: *A One-Equation Turbulence Model for Aerodynamic Flows.* AIAA 1992-0439, 1992.

[89] SPURK, JOSEPH H.: *Strömungslehre.* Springer-Verlag, Berlin, 1996.

[90] STEFFENS K., SCHÄFFLER A.: *Triebwerksverdichter - Schlüsseltechnologie für den Erfolg bei Luftfahrtantrieben.* DGLR-Jahreskongress IDJT2000-001, 2000.

[91] TAKEISHI K., MATSUURA M., AOKI S., SATO T.: *An experimental study of heat transfer and film cooling on low aspect ratio turbine nozzles.* ASME Paper GT1989-187, 1989.

[92] TORRE D., VAZQUEZ R., DE LA ROSA BLANCO E., HODSON W.: *A new Alternative for Reduction of Secondary Flow in Low Pressure Turbines.* ASME Paper GT2006-91002, 2006.

[93] VANDERPLAATS, G.N.: *Numerical Optimisation Techniques for Engineering Design.* McGraw-Hill, 1984.

[94] WADIA A.R., SZUCS P.N., CRALL W.W.: *Inner Workings of Aerodynamic Sweep.* ASME GT1997-401, 1991.

[95] WEISS A.P., FOTTNER L.: *The Influence of Load Distribution on Secondary Flow in Straight Turbine Cascades.* Journal of Turbomachinery, Vol. 117:133–141, 1995.

[96] ZHONG J., HAN J., LIU Y., TIAN F.: *Numerical Simulation of Endwall Fence on the Secondary Flow in Compressor Cascade.* ASME GT2008-50888, 2008.

A Appendix

A.1 Derivation of the Total Pressure Loss Coefficient for Rotating Blade Rows

In the following, a brief derivation of the total pressure loss coefficient for rotating blade rows will be given. In rotating blade rows, a common quantity used to express the energy content of the fluid is known as the rothalpy.

$$I = h + \frac{w^2}{2} - \frac{U^2}{2} \tag{A.1}$$

Rothalpy has analogous properties in rotating blade rows compared to the stagnation enthalpy in stationary passages. According to Cumpsty [9], the rothalpy is constant in a moving passage under certain conditions, such as the flow is steady in the rotating frame, no work is introduced into the flow in the rotating frame and there is no heat transfer occuring. Then, the rothalpy at the blade inlet and blade outlet should be the same. If we consider the first two terms on the right hand side of equation A.1 as the total enthalpy

$$h_{t\_rel} = h + \frac{w^2}{2} = c_p T_{rel} \tag{A.2}$$

and describe the entrainement velocity by

$$\frac{U^2}{2} = \frac{(r\omega)^2}{2} \tag{A.3}$$

we can express the constraint of the rothalpy remaining constant between inlet and outlet of a moving blade row, i.e. $I_{in} = I_{out}$, by

$$c_p T_{rel\_in} + \frac{(r_{in}\omega)^2}{2} = c_p T_{rel\_out} + \frac{(r_{out}\omega)^2}{2}. \tag{A.4}$$

In an axial compressor, where the streamlines enter and leave the blade row on the same radii, the total relative temperature would have to remain constant if the mentioned conditions applied [9]. However, if the streamlines enter and leave the blade row on different radii, e.g. due to merging or diverging end walls as it is the case for the investigated compressor, the modification of the streamline radii between inlet and outlet alone leads to an according change in total relative temperature pressure due to the change of the local blade speed. The matching radial positions at inlet and outlet have to be evaluated according to the corresponding relative channel height to consider the same streamline. Then, equation A.4 can be used to determine this change in total pressure across the blade for the ideal case without any losses leading to the new isentropic temperature T' at the outlet

$$T'_{rel\_out} = T_{rel\_in} + \frac{\omega^2}{2c_p} \left(r_{out}^2 - r_{in}^2 \right). \tag{A.5}$$

Using the isentropic relation, we can calculated the corresponding total pressure in the relative system the fluid would have at the outlet due to the change in radius to

$$p'_{rel\_out} = p_{rel\_in} \left(\frac{T'_{rel\_out}}{T_{rel\_in}} \right)^{\left(\frac{\gamma}{\gamma-1} \right)}. \tag{A.6}$$

p'_{rel_out} can be considered as the rothalpic isentropic pressure at rotor outlet that would appear in the ideal case without any losses. Relating this pressure to the real value of the total relative pressure at this radial position leads to the total pressure loss coefficient for rotating blade rows given by

$$CP_{rel} = \frac{p'_{rel\_out}(r) - p_{rel\_out}(r)}{p_{dyn\_in}(r)}. \tag{A.7}$$

A.2 Rotor Wake Influence on the Datum Stator Blade at Near Stall

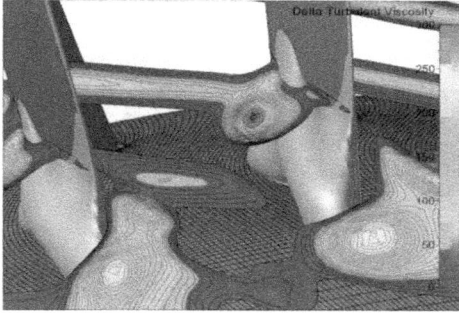

Figure A.1: Rotor wake near stall filtered by TAVG solution at time step $t = 2$

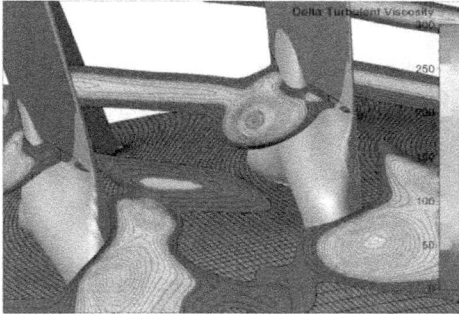

Figure A.2: Rotor wake near stall filtered by TAVG solution at time step $t = 4$

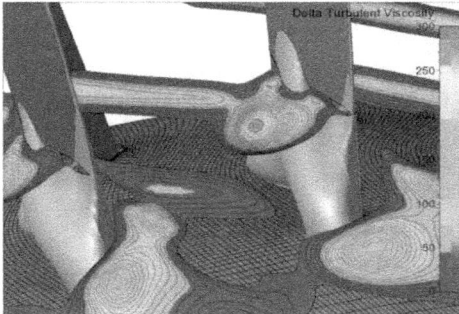

Figure A.3: Rotor wake near stall filtered by TAVG solution at time step $t = 6$

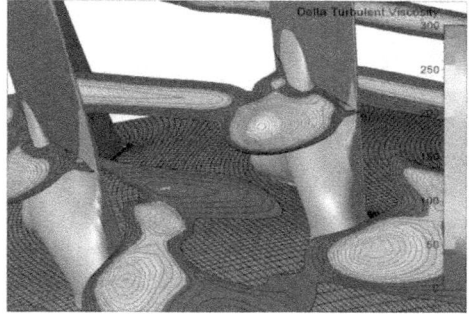

Figure A.4: Rotor wake near stall filtered by TAVG solution at time step $t = 8$

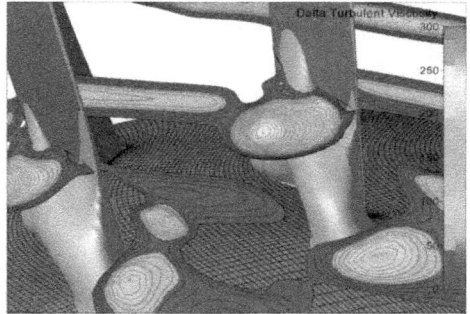

Figure A.5: Rotor wake near stall filtered by TAVG solution at time step $t = 10$

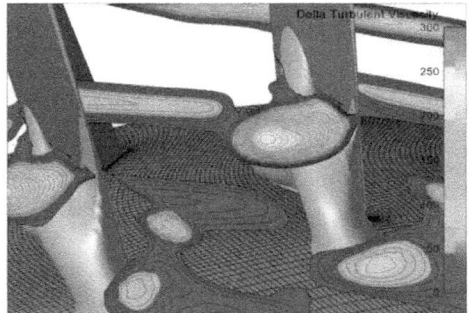

Figure A.6: Rotor wake near stall filtered by TAVG solution at time step $t = 12$

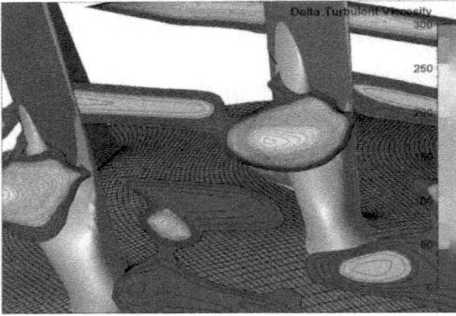

Figure A.7: Rotor wake near stall filtered by TAVG solution at time step $t = 14$

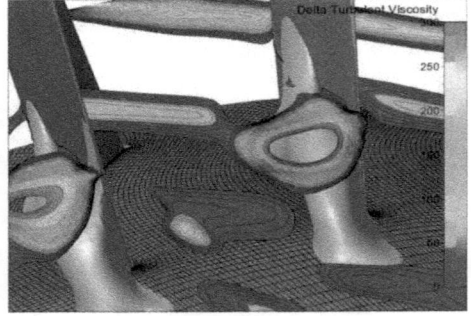

Figure A.10: Rotor wake near stall filtered by TAVG solution at time step $t = 20$

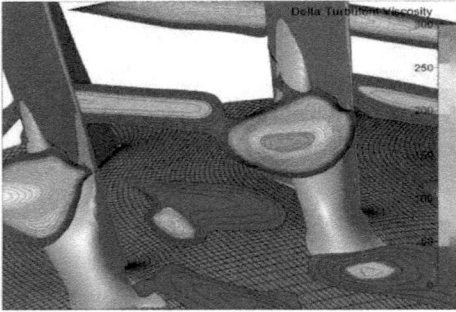

Figure A.8: Rotor wake near stall filtered by TAVG solution at time step $t = 16$

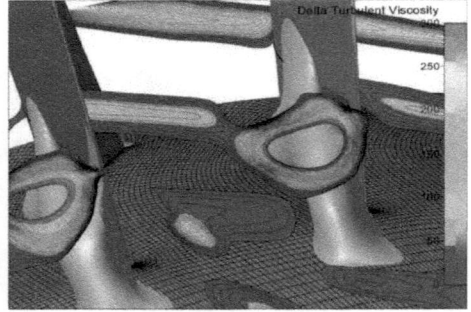

Figure A.11: Rotor wake near stall filtered by TAVG solution at time step $t = 22$

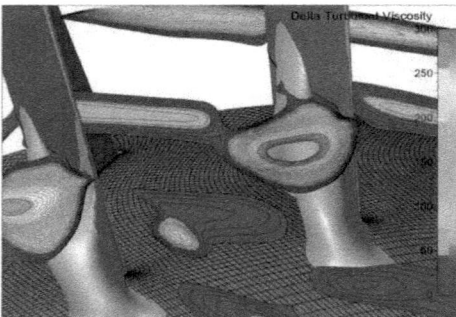

Figure A.9: Rotor wake near stall filtered by TAVG solution at time step $t = 18$

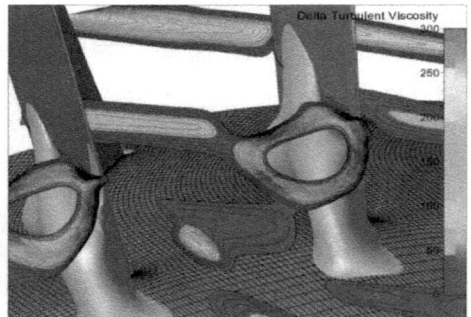

Figure A.12: Rotor wake near stall filtered by TAVG solution at time step $t = 24$

A Appendix

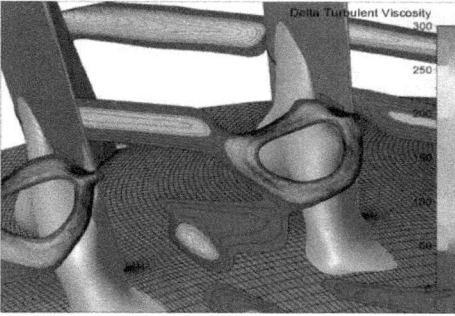

Figure A.13: Rotor wake near stall filtered by TAVG solution at time step $t = 26$

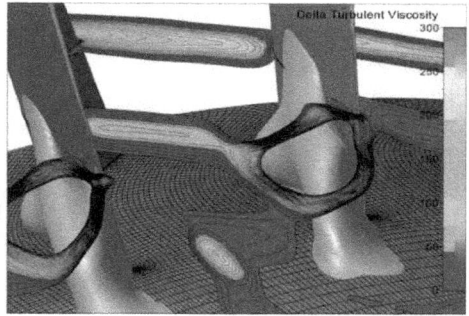

Figure A.16: Rotor wake near stall filtered by TAVG solution at time step $t = 32$

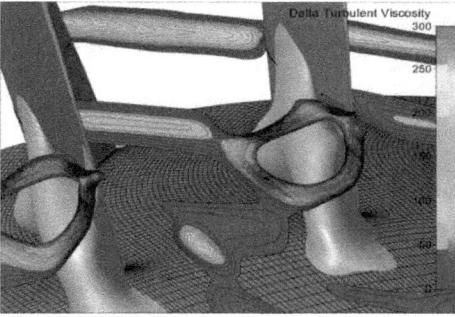

Figure A.14: Rotor wake near stall filtered by TAVG solution at time step $t = 28$

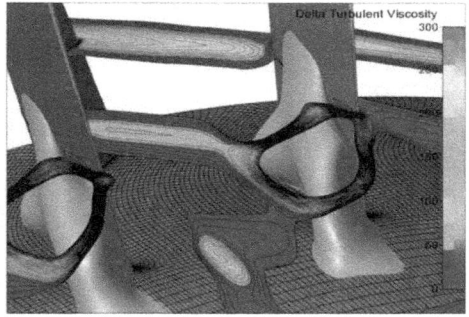

Figure A.17: Rotor wake near stall filtered by TAVG solution at time step $t = 34$

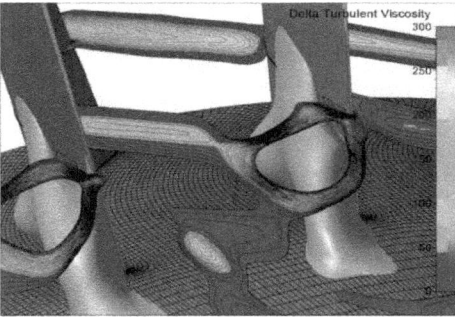

Figure A.15: Rotor wake near stall filtered by TAVG solution at time step $t = 30$

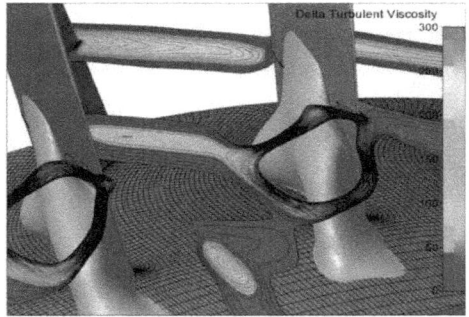

Figure A.18: Rotor wake near stall filtered by TAVG solution at time step $t = 36$

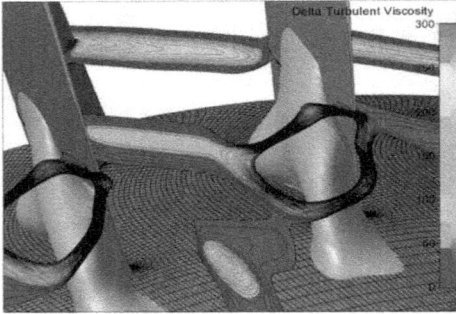

Figure A.19: Rotor wake near stall filtered by TAVG solution at time step $t = 38$

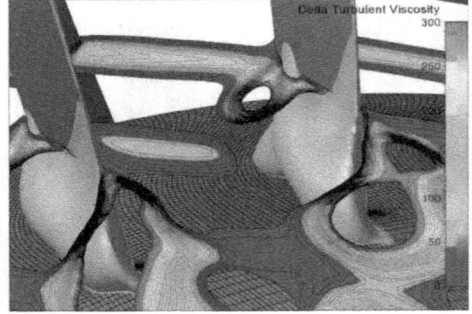

Figure A.22: Rotor wake near stall filtered by TAVG solution at time step $t = 44$

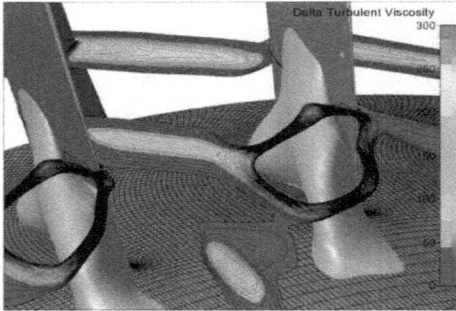

Figure A.20: Rotor wake near stall filtered by TAVG solution at time step $t = 40$

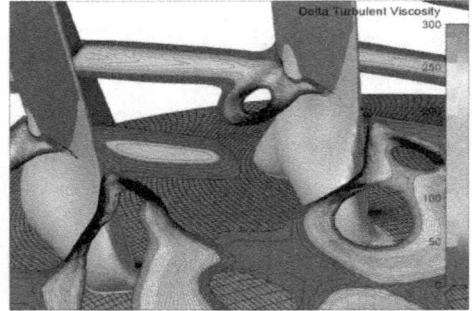

Figure A.23: Rotor wake near stall filtered by TAVG solution at time step $t = 46$

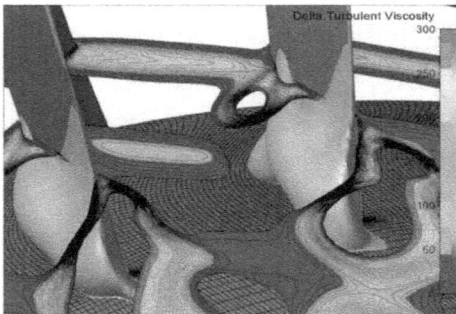

Figure A.21: Rotor wake near stall filtered by TAVG solution at time step $t = 42$

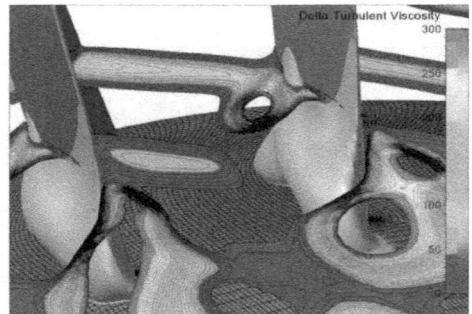

Figure A.24: Rotor wake near stall filtered by TAVG solution at time step $t = 48$

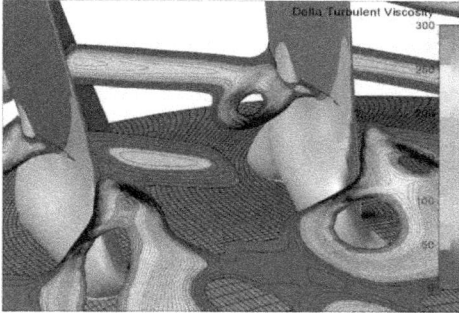

Figure A.25: Rotor wake near stall filtered by TAVG solution at time step $t = 50$

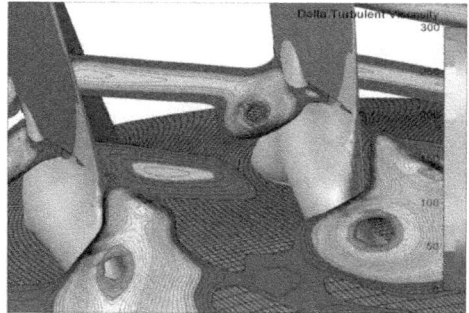

Figure A.28: Rotor wake near stall filtered by TAVG solution at time step $t = 56$

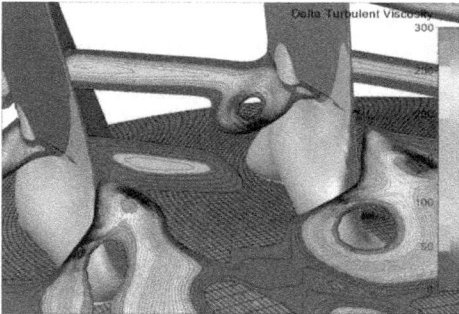

Figure A.26: Rotor wake near stall filtered by TAVG solution at time step $t = 52$

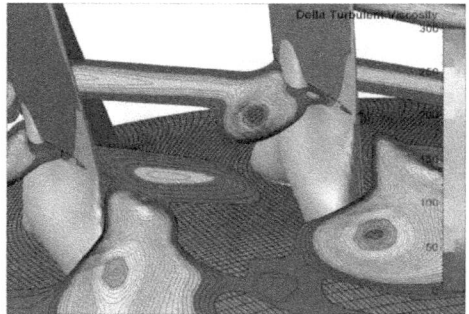

Figure A.29: Rotor wake near stall filtered by TAVG solution at time step $t = 58$

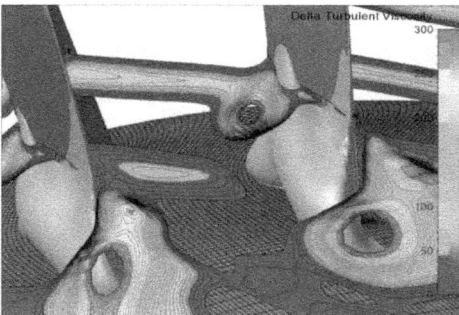

Figure A.27: Rotor wake near stall filtered by TAVG solution at time step $t = 54$

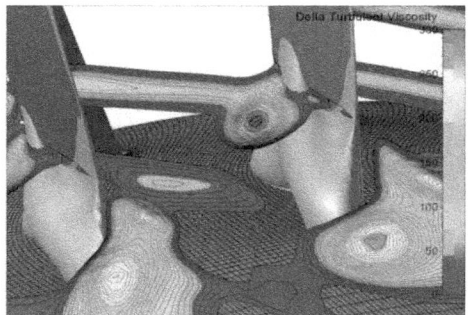

Figure A.30: Rotor wake near stall filtered by TAVG solution at time step $t = 60$

A.3 Rotor Wake Influence on the Datum Stator Blade at Design Conditions

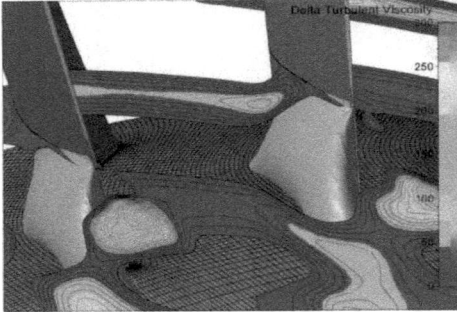

Figure A.31: Rotor wake at design conditions filtered by TAVG solution at time step $t = 4$

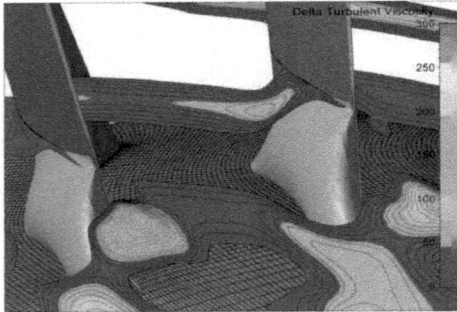

Figure A.32: Rotor wake at design conditions filtered by TAVG solution at time step $t = 8$

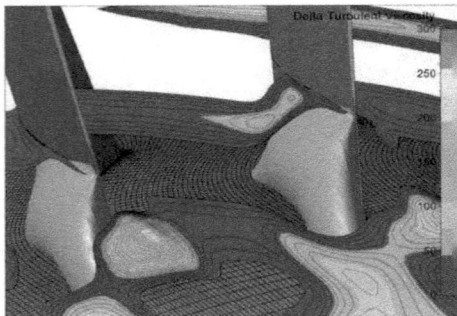

Figure A.33: Rotor wake at design conditions filtered by TAVG solution at time step $t = 12$

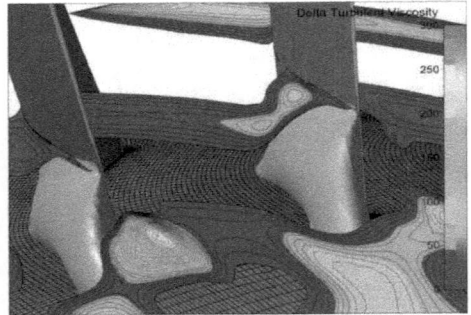

Figure A.34: Rotor wake at design conditions filtered by TAVG solution at time step $t = 16$

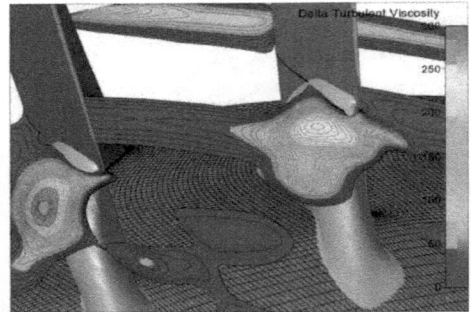

Figure A.35: Rotor wake at design conditions filtered by TAVG solution at time step $t = 20$

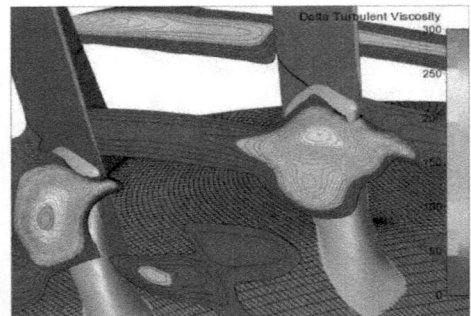

Figure A.36: Rotor wake at design conditions filtered by TAVG solution at time step $t = 24$

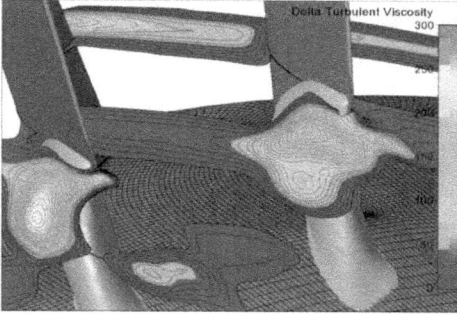

Figure A.37: Rotor wake at design conditions filtered by TAVG solution at time step $t = 28$

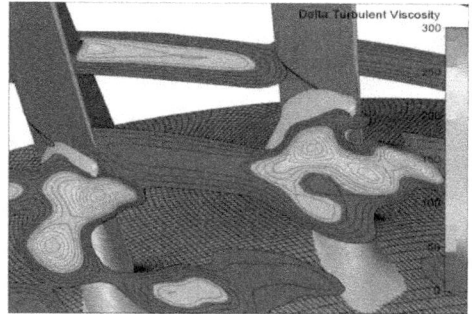

Figure A.40: Rotor wake at design conditions filtered by TAVG solution at time step $t = 40$

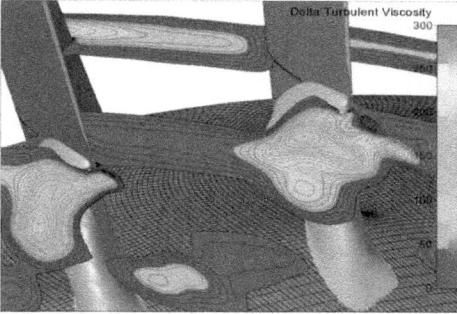

Figure A.38: Rotor wake at design conditions filtered by TAVG solution at time step $t = 32$

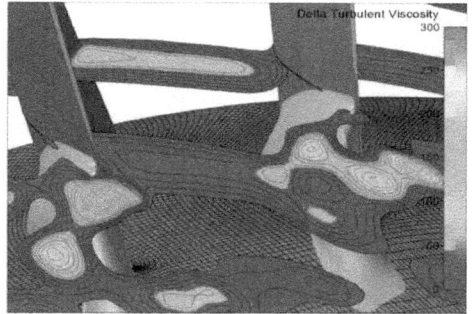

Figure A.41: Rotor wake at design conditions filtered by TAVG solution at time step $t = 44$

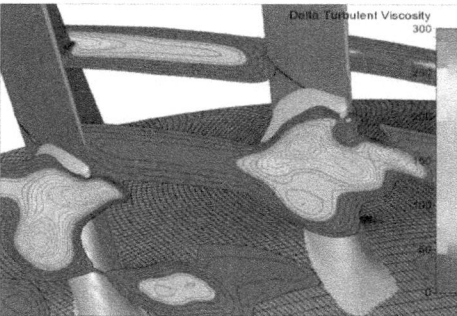

Figure A.39: Rotor wake at design conditions filtered by TAVG solution at time step $t = 36$

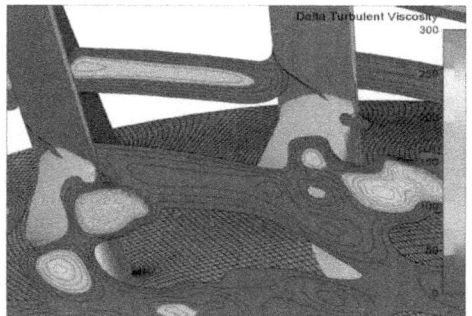

Figure A.42: Rotor wake at design conditions filtered by TAVG solution at time step $t = 48$

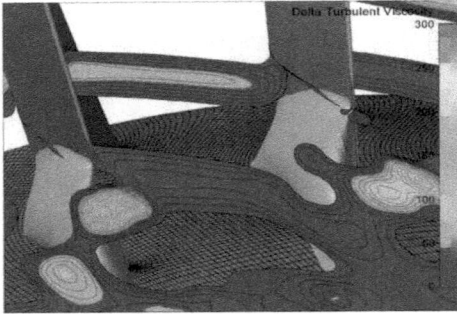

Figure A.43: Rotor wake at design conditions filtered by TAVG solution at time step $t = 52$

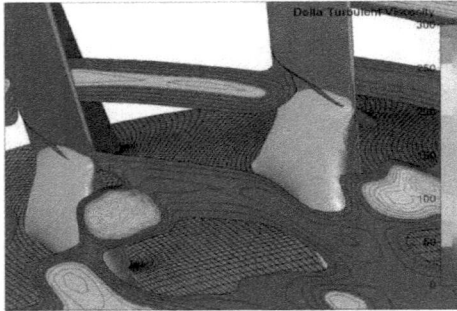

Figure A.44: Rotor wake at design conditions filtered by TAVG solution at time step $t = 56$

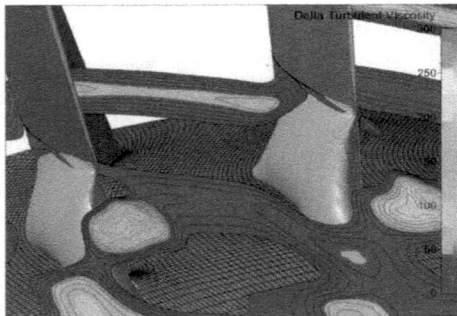

Figure A.45: Rotor wake at design conditions filtered by TAVG solution at time step $t = 60$